Business Organisation for Construction

Considerable skill is needed to operate a business in today's construction industry. Organisations are often large and complex, and relationships both internally and with one another are not easily understood.

Beginning with an overview of key management thinkers and theorists, Chris March shows how businesses are set up, organised and planned in the context of the special circumstances of the construction industry. In a sector which employs many hundreds of thousands, it is vital to understand how people are organised, recruited, looked after, motivated and led. This book gives a clear insight into human resource management in the industry, and explains how people negotiate and communicate with each other, in terms of personal, group and technological support.

Chris March has a wealth of practical experience in both the construction industry and teaching students. His down-to-earth approach and mixture of theory and real-life evidence from personal experience make this an unequalled insight into how business organisations in the construction industry really work.

Chris March is a graduate from Manchester University. He worked for John Laing Construction and later for John Laing Concrete where he became Factory Manager. On entering higher education he worked in both the UK and Hong Kong before joining the University of Salford becoming Senior Lecturer and then the Dean of the Faculty of the Environment. He is a former winner of the Council for Higher Education Construction Industry Partnership Award for Innovation.

Business Organisation
of Construction

Business Organisation for Construction

Chris March

Spon Press
an imprint of Taylor & Francis
LONDON AND NEW YORK

First published 2009
by Taylor & Francis
2 Park Square, Milton Park, Abingdon, Oxon OX14 4RN

Simultaneously published in the USA and Canada
by Taylor & Francis
270 Madison Ave, New York, NY 10016

Taylor & Francis is an imprint of the Taylor & Francis Group, an informa business

Typeset in Sabon by
HWA Text and Data Management, London
Printed and bound in Great Britain by
TJ International Ltd, Padstow, Cornwall

British Library Cataloguing in Publication Data
A catalogue record for this book is available from the British Library

Library of Congress Cataloging-in-Publication Data
March, Chris.
Business Organisation for Construction / Chris March.
p. cm.
Includes bibliographical references and index.
1. Construction industry – Management. I. Title.
TH438.M3067 2009
690.068´1–dc22 2008037920

ISBN10: 0–415–37009–4 (hbk)
ISBN10: 0–415–37010–8 (pbk)
ISBN10: 0–203–92802–4 (ebk)

ISBN13: 978–0–415–37009–7 (hbk)
ISBN13: 978–0–415–37010–3 (pbk)
ISBN13: 978–0–203–92802–8 (ebk)

Contents

Figures

Tables

Introduction

This book is one of three closely related texts, *Finance and Control for Construction*, *Operations Management for Construction* and *Business Organisation for Construction*; the reason for writing these books is the increasing awareness of the shortage of new texts covering the whole range of construction management. There are plenty of good recent texts appropriate for primarily final-year and postgraduate students, but they tend to be subject specific and assume a certain level of knowledge from the reader. It also means students find this cost prohibitive and tend to rely upon the library for access. (The research selectivity exercises have encouraged authors to write books based upon their research, for which credit has been given in the assessment, whereas none has been given to those writing textbooks.) The purpose of these three books is an attempt to give students the management vocabulary and understanding to derive greater value from these specialist texts.

The original intention was to write this with construction management undergraduate students in mind, but as the project developed it became clear that much of the subject matter was appropriate for all the construction disciplines. In more recent times the industry, being undersupplied with good construction graduates, has turned to recruiting non-cognate degree holders and many of these are and will study on Masters courses in construction management. These texts are ideally suited to them as background reading in giving a broad base of understanding about the industry.

Rather than having a large number of references and bibliographies at the end of each chapter, generally I have limited these to a few well-established texts, some referenced in more than one chapter, so the reader is directed to only a few if wishing to read further and in more depth on the subject. The chapters vary in length considerably depending upon the amount of information I believe is relevant at this level.

The aim of this book (the third of three) is to introduce the reader to key management thinkers, the way businesses are formed and organised, including strategic planning, the management of change and marketing. Human resources are explored and what makes good teams and leaders and how to manage stress. There is a chapter on risk management, which summarises risks mentioned in all three texts. Finally the subject of communications is considered both at a personal and corporate level.

Of the two related books – *Finance and Control for Construction* is concerned with tracking through each stage of the process the control of finance, with consideration taken of sustainability and the environment; and *Operations Management for Construction* is concerned primarily with the site activity production issues, management of suppliers and sub-contractors, and quality assurance. There is inevitably overlap in all three books, so I have cross-referenced from one book to another and within each subject, with the hope of aiding readers.

On a personal note I believe that there is no definitive way of managing and, as Mike Stoney, the Managing Director of Laing, used to say to students, 'Don't copy me, it may not suit your personality, but watch and listen to other successful mangers and pinch the bits from them that suit you.' I totally agree and, for what it is worth, I have added some other comments and thoughts of other people that have influenced my way of thinking over the years.

My head master, Albert Sackett, who taught me to assume everything was wrong until I could prove it correct. He would set an essay on say 'define the difference between wit and humour'. After he had marked it he would sit you in front of the class and then debate with you your answer. Having convinced you he was right and you were wrong, would then reverse role and argue back the opposite way.

Godfrey Bradman, Chairman of Rosehaugh plc and now Chairman of Bradman Management Services, reinforced my views from school of not simply accepting any thing you are told, but, in his case, also to have the ability to ask the right questions, usually simple ones such as 'why not?'

My father who taught me to accept failure was a fact of life and not to hide the fact, but to accept and admit it and get on with life having learnt from the experience. I also have found that by admitting it, 'the punishment' was always less than being found out. When in the precast factory is was always easier to advise the site that the load of components was going to be late or not delivered that day than await the angry phone call demanding to know what had happened. It also made sense because, although disappointed, they had more time to rearrange their own schedule of work.

Dorothy Lee, retired Deputy Director of Social Services in Hong Kong responsible for the Caritas operated Kai Tak East Vietnamese refugee camp who advised the small group I led in developing a self-build solution for refugees, which after many weeks of hard work was no longer required, said 'I know you will be disappointed, but remember you have grown a little more as a result.'

John Ridgway, explorer and outside activities course organiser, had at his School of Adventure, based in Sutherland, Scotland, the adage of positive thinking, self-reliance and to leave people and things better than you find them. He also made a very clear impression on me of the importance when in charge, to have the ability to stand outside the circle and view the problem from outside and allocate tasks without becoming too closely involved.

Don Stradling, the Personnel Director of Laing and senior negotiator with the Federation of Civil Engineers Employees with the very simple piece of advice that 'you should always keep the moral high ground'. How right he is. It is surprising the number of people that don't, and when confronted with one that does they almost invariable fail in the negotiation. It also results in having respect from those they have contact with, as they believe in your integrity and accept what you say is meant in an honourable way.

Finally Dennis Bate, member of the main board of Bovis Lend Lease who told me that he, throughout his life from leaving school at sixteen to become an apprentice joiner, strove to do whatever he did to the best of his ability and better than anybody else.

I wish to acknowledge the support and help given by so many in putting together these three books. From the construction industry, staff from Bovis Lend Lease, Laing O'Rourke and Interserve in particular, who have spent many hours discussing issues and giving advice. It was at Plymouth University the idea to produce these texts was formulated and where colleagues gave me encouragement to commence, and then when at Coventry University, not only was this continued, but also doors were always open whenever I wished to consult on an idea or problem. Having spent so many years at Salford, the years from 1987 onwards, when the we started the Construction Management degree, were of great significance in developing ideas on the needs of construction management students and this would not have been possible without the assistance and guidance from my colleagues there at the time, especially Tony Hills, John Hinks and Andy Turner, as well as the many supporting contractors always on hand to give advice, ideas and permit access to other colleagues in their organisations. Finally, to my wife Margaret who has to suffer many hours on her own whilst I locked myself away in the study, but never ceased to give her support and encouragement throughout.

Chris March

CHAPTER

Pioneers of management theory

1.1 Definitions

There are terms regularly used in management 'speak' so the author has given his own definitions based on his own readings and experiences.

Management: Each situation is almost invariably unique. Management should assess all the relevant factors available and make a decision concerning the problem in question.

Administration: Administrators work primarily to a set of rules and decisions are made based on these guidelines. There can be some flexibility in 'bending the rules' but usually administrators are concerned that a new precedent might be set if the rules are not followed. It should be noted that in most organisations, especially medium and large, management and administration functions exist side by side. For example, wages and salaries would be seen as an administrative function.

Accountability: An employee is accountable for his or her actions and reports to a supervisor.

Responsibility: An employee is responsible for actions taken, but is also responsible for the actions of those reporting directly to them. Responsibility registers highly in employee job satisfaction.

Delegation: Is probably the hardest act to carry out by a manager. In essence it means delegating part of one's responsibility to a subordinate and allowing that person to have complete responsibility for the work delegated. Many managers fall down on this by either delegating roles and jobs they do not wish to carry out, or worse still, continue to take actions and decisions relating to the tasks delegated. This latter failure is often due to a lack of confidence in others. It can be extremely frustrating for the subordinate.

1.2 Introduction

There are far too many contributors to management theory to cover all, so those selected are people who have made, in the author's opinion, significant contributions and to whom he can closely relate, relative to his own experiences. This is important for the reader to understand, as there are many others who have contributed equally and may well be more in tune with the experiences of the reader. The author has also 'cherry picked' the topics that have influenced him and these in no way represent a complete coverage of the respective writers' works. Hence when browsing through the many management texts on offer, readers should widen their knowledge beyond the content of this book.

1.3 Development of management theory

Serious management thinking has been developing for just over 100 years. It can be broadly classified into four categories: classical, human relations, systems and contingency, which have occurred sequentially as approximately demonstrated in Figure 1.1. It is difficult to indicate precisely when one finishes and another starts. Pre-classical management was based primarily on trial and error, but the classical school saw the need to apply scientific approaches to management, at the exclusion of the human factor. This exclusion was addressed at the next phase when the human element was considered. After the Second World War the theory was redirected to look at systems and processes. More recently there has been awareness that all schools had something to offer and a more holistic approach has been taken.

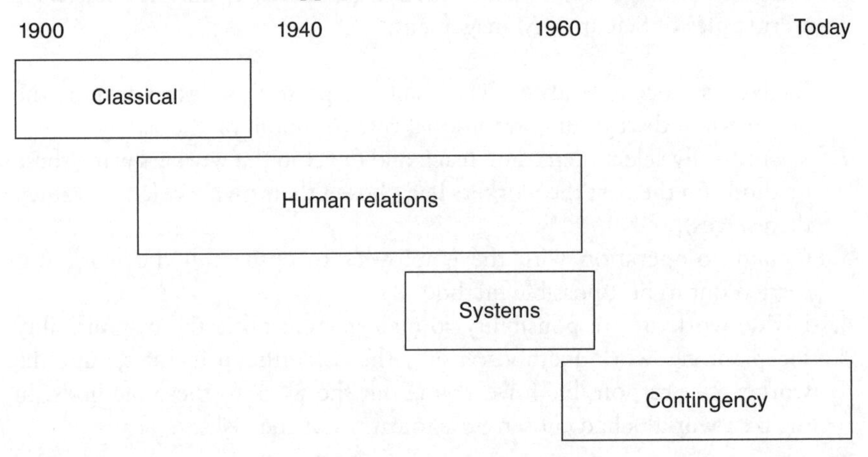

Figure 1.1 Timeline of management theory schools

1.4 The classical school

It is necessary to place this development of management thought into its historical context. At the end of the nineteenth century labour was cheap, plentiful and considered primarily as a resource, rather than a 'human resource'. If the steel furnaces were not being filled with coal fast enough, you just added extra manpower. Little or no thought was placed on the efficiency of labour. There are three key theories in this school: Scientific, Administrative and Bureaucratic, developed primarily by Taylor, Fayol and Weber respectively.

1.4.1 Scientific management – F.W. Taylor (1856–1915)

Taylor joined the Midvale Steel Works in Pennsylvania, USA, as a machine shop labourer in 1878 and, within six years, had risen through the ranks to become chief engineer after which he moved to the Bethlehem Steel Works. He observed that workers were not working at their full potential and came to the conclusion it was because they were concerned that if they worked flat out, others would lose their jobs. The wage systems in place discouraged higher productivity, as they were often structured so if productivity increased above the set standard, incentive payments would be cut. Many of the working methods had been handed down from generation to generation of employees without any consideration on how to improve productivity.

Taylor was interested in applying scientific techniques to management believing this was the way to deal with low productivity and developed his Four Principles of Scientific Management:

1. Evolve a science to study each element of a person's work to develop the best method disregarding traditional rule of thumb methods.
2. Scientifically select then train, teach and develop the workers using these methods (in the past the workers had chosen their own work and trained themselves).
3. Obtain co-operation with the employees to ensure all the work was carried out to best possible method.
4. Divide work and responsibility so management takes the responsibility for planning work methods, using the scientific principles, and the workers are responsible for carrying out the work to these methods. In the past workers had chosen how to carry out the tasks.

Taylor was interested in measuring the output of operatives to establish the best way to carry out tasks which he called time study, subsequently referred to as time and motion study which became the basis of work study as known today (*Operations Management for Construction,* Chapter 3). He also developed wage incentive schemes to increase productivity.

One of his famous studies was the use of shovels in the Bethlehem Steel Works. He noted that operatives used the same shovel irrespective of the type and weight of the material they were loading. Using scientific measurement he established what mattered was the total weight of both the shovel and the materials, which he determined to be 21 pounds. He was able to increase shovelling productivity almost four-fold as a result.

1.4.2. Administrative management – Henri Fayol (1841–1925)

Whilst Taylor was developing his ideas in America, Henri Fayol was revolutionising management thinking in France. He was brought up in a upper middle-class family and trained as a mining engineer, working his way up to become managing director of Commentry-Fourchampboult et Decazeville, a coal mining and foundry combine. Whereas much of Taylor's work concentrated on the production part of the process, Fayol was interested in management principles from the chief executive's point of view. He was concerned with designing an efficient organisation structure based on administration.

He developed a definition of management based on five functions (note some texts refer to six, separating forecasting and planning), which in essence are still valid. These are:

- forecast and plan
- organise
- command
- co-ordinate
- control.

Any business needs to examine and forecast what is likely to happen in the future, be it the marketplace in which one is to sell products or services, i.e. the demand side, or on the supply side to establish the availability of resources, notably labour and materials. All of this takes account of economic trends. The company has to produce a corporate plan that brings together the needs of all sections of the business to meet the corporate objectives, followed by operational plans to assist in executing these needs in practice. These should be flexible enough to take account of changing circumstances and assist in predicting courses of action.

The management needs to provide the material and human resources and build up an organisational structure so the work can be executed to meet the plan. This requires management to stand back and decide on a structure; in other words, how the business should be divided into groups of activities or functions, who should be responsible to whom, the number of people one should be responsible for and so on.

Logically following on from the above is the need to have the authority to command so that the organisation, now structured, can enable optimum return from all employees in meeting the corporative objectives. Command is the relationship between a manager and his or her immediate subordinates. The quality of command results from the manager's ability to demonstrate knowledge of the business, subordinates, the quality of contact with them, and to inspire confidence.

However, this in itself will be insufficient if the interface between the different parts of the organisation is not explored and co-ordinated. Fayol wrote that co-ordination is necessary in 'binding together, unifying and harmonizing all activity and effort'. Too often departments work in isolation without understanding how their actions impact on another. This issue is addressed in ISO 9001:2004 Quality Management Systems Requirements written a century later (*Operations Management for Construction*, Chapter 8).

The final element is to monitor that all the previous elements are functioning in line with the plans, commands and instructions.

He further expounded 14 principles as follows:

1. *Division of work*
   By performing only one part of the job, a worker can produce more and better work for the same effort. Specialisation is the most efficient way to use human effort.
   This view was the basis of much of the thinking in mass production, such as cars, where each worker was given a clearly defined task.
2. *Authority and responsibility*
   Authority is the right to give orders and obtain obedience, and responsibility is the corollary of authority.
   This ignores the human element that today might question the competence of the superior requiring that authority should also be linked to the need to obtain respect. However at the time the position automatically commanded respect and authority.

3. *Discipline*
Obedience to organisational rules is necessary.
The best way to have good superiors, and clear and fair rules and agreements is to apply sanctions and penalties judiciously. Nobody would disagree with this in principle, but the emphasis is on punishment rather than on congratulations and positive motivation. There is a danger that by taking this too literally there becomes a conflict with 14 below.
4. *Unity of command*
There should be one and only one superior for each individual employee.
This on the face of it is correct especially in the context of executive line management, but as can be seen in functional management organisations (section 2.3.2), there are legitimate situations where a person is responsible to two people, one for discipline and performance, and the other for the quality of their expertise.
5. *Unity of direction*
All units in the organisation should be moving towards the same objectives through co-ordinated and focused effort.
This would be a valid statement today.
6. *Subordination of individual interest to general interest*
The interests of the organisation should take priority over the interests of an individual employee.
This should occur at all levels of the organisation. Whilst the concept is valid, there can be conflicts with this and the ambitions of employees who can often usurp this position by providing their superior with the views and answers they think they want, so as to impress and be noticed, whereas a healthy debate and disagreement may better serve the organisation.
7. *Remuneration of employees*
The overall pay and compensation for employees should be fair to both the employees and the organisation.
Any organisation needs to balance what it can afford with an equitable distribution to its employees. There is no perfect system even today as the endless debates over 'fat cats' and pay compatibility demonstrates.
8. *Centralisation*
There should be a balance between subordinates' involvement through decentralisation and managers' retention of final authority through centralisation.
Centralisation is always present to a greater or lesser extent depending on the size of the company and quality of its managers. The larger the organisation, the more stretched are the lines of communication between

the board and the lower parts of the organisation. This makes control more difficult as the further away from the centre the more autonomous the parts can become. How much is permitted should be a function of the capabilities of the personnel.

9. *Scalar chain*
   Organisations should have a train of authority and communication that runs from the top to the bottom and should be followed by managers and subordinates.
   A follow-on from centralisation. The longer the chain of the command is, the more difficult it is for the top management to know what is going on at the bottom and they become cut off. On the other hand, if these lines of communication are reduced, the top manager has many more people to control which in turn can cause difficulties. Lateral communication is encouraged providing the vertical management chain is kept informed.
10. *Order*
    People and materials must be in suitable places at the appropriate time for maximum efficiency; that is, a place for everything and everything in its place.
    No change here; indeed this might be used as the starting point for just in time deliveries (*Operations Management for Construction,* Chapter 6).
11. *Equity*
    Good sense and experience are needed to ensure fairness to all employees who should be treated as equally as possible.
    A law of management which most strive to achieve.
12. *Stability of personnel*
    Employee turnover should be minimised to maintain organisational efficiency.
    No argument with this.
13. *Initiative*
    Workers should be encouraged to develop and carry out plans for improvement.
    A pre-runner of McGregor's Y theory (section 1.5.5).
14. *Esprit de corps*
    Management should promote a team spirit of unity and harmony among employees.
    Many construction companies have used this philosophy in naming their in-house publications such as Taylor Woodrow's 'Pulling together' and Laing's 'Team Spirit'. However, to create this unity and harmony requires a deep understanding of motivation theory and co-operation of all levels of management.

1.4.3 Bureaucratic Management – Max Weber (1864–1920)

The German sociologist, Max Weber, believed organisations should operate on a rational basis and not on the arbitrary decisions made by the owners, which were usually based on nepotism. He believed this to be not only unfair, but also a waste of talent. He advanced the concept of bureaucracy. Today the term bureaucracy has connotations of red tape and excessive numbers of rules and regulations, but Weber used it in the context of the way authority was exercised within organisations. He distinguished power as being the ability to make people obey regardless of their resistance, and authority as commands being obeyed voluntarily. He recognised three pure types of legitimate authority:

1. *Charismatic*. The authority based on leadership qualities or strength of personality. There are many examples of these throughout history, in politics, the military and religion, but especially since the industrial revolution, in industry as well. The problem arises when considering succession, as the 'groomed' replacement may not have the same charismatic personality. Indeed in such an event history is littered with examples of rivalry between potential successors.
2. *Traditional*. The type of authority based on custom, tradition and precedence. An example of this is hereditary authority when the son and heir takes over, with everybody expecting and assuming this will happen. Another form is patronage, the giving of position for example by the monarch or prime minister.
3. *Rational/legal*. It is called rational because the means are designed to achieve specific goals. Personnel are selected because of their ability and they have a clear understanding of their part in the overall process, each part working as if in a well-oiled machine. The legal authority is based on their position in the hierarchy at that time using established rules and procedures. The individual only has the authority whilst in post and ceases to have that specific authority if moved elsewhere in the organisation. Weber believed that a bureaucratic organisation as he defined it, was the best.

His organisational structure has the following characteristics:

- tasks are divided into very specialised jobs;
- a rigorous set of rules must be followed to ensure predictability and eliminate uncertainty in task performance;

- there are clear authority–responsibility relationships that must be maintained;
- superiors take an impersonal attitude in dealing with subordinates;
- employment and promotions are based on merit;
- lifelong employment is an accepted fact.

These types of organisations still exist, but tend to be rigid, inflexible and heavily reliant on red tape and it is often difficult to identify clearly the more capable employees.

1.4.4 Others of interest

The reader should also be aware of the work of Henry Gantt (1861–1919) noted for the 'Gantt chart', these days referred to as the bar chart, and his development of incentive schemes for supervisors which were linked to the output and incentives paid to their workers. Frank Gilbreth (1868–1924) was interested in time and motion studies and efficiency of human movement. He did a considerable amount of work on the bricklaying process and, by designing appropriate scaffolding and stipulating the consistency of mortar, was able to almost treble output. His wife Lillian (1878–1972) worked with him and was also interested in the human aspects of work such as the selection, placement and training of personnel.

1.5 Human relations school (behavioural sciences)

This school of thought emerged as workers became more organised and managers sought to achieve increased productivity and man-management. It was in part a reaction to the rigidity of the classical school and an awareness of manpower being a more complex resource than had been previously been considered. The key theories are: Group Work, Democratic Decisions, Motivation and Participation.

There is often considerable overlap between these so it is not proposed to discuss them individually, but to look at the contributions various pioneers made and allow the reader to distinguish accordingly.

1.5.1 Elton Mayo (1880–1949)

Elton Mayo was an Australian academic who spent most of his working life at Harvard University and is credited by most as being the founder of both the human relations movement and industrial sociology. The behavioural

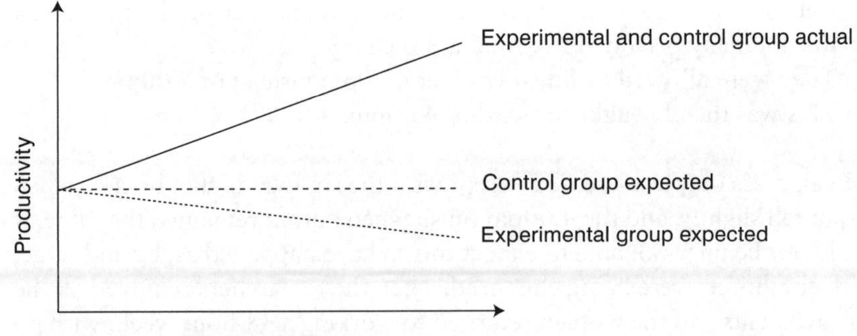

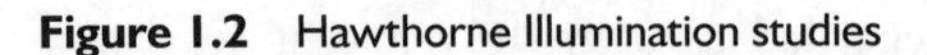
Figure 1.2 Hawthorne Illumination studies

approach is generally considered to have started with the experiments carried out at the Hawthorne plant of Western Electric Company (1924, 1927–1932). These are referred to as the Hawthorne Studies. In the first studies, called the 'Illumination studies', the control group worked with constant light, whereas the experimental group had the lighting steadily reduced until such time as the workers in the group complained there was inadequate light to work. At this point output began to decline in the experimental group, but surprisingly up until then production had risen almost identically in both groups, as demonstrated in Figure 1.2. The researchers deduced that something other than the levels of lighting was affecting performance.

This unexpected result needed further resolution and between 1927 and 1932, Mayo led a team of researchers to explore the matter further in what was to become known as the Relay Assembly Test Room experiment. A group of six women employed to assemble telephone relays were separated from the rest of the workforce so their output and morale could be observed when changes were made to the working arrangements. The normal working week at the start of the experiment was a 48-hour week with no rest breaks and included working on Saturdays. Over a period of five years the following changes were made:

- A special group incentive scheme was introduced. Prior to this they had been grouped together with a hundred other operatives for incentive purposes.
- Two five-minute rest periods were introduced, one in the morning the other in the afternoon.
- These rest periods were increased to ten minutes.
- Six five-minute rest periods were introduced, but the women complained their work patterns were broken up by so many breaks.

- The two ten-minute breaks were introduced and during the first break the company provided hot food at no charge.
- They were allowed to finish work at 4.30pm instead of 5.00pm.
- This was then brought forward to 4.00pm.

In all cases output increased except with the six five-minute breaks, when output fell slightly, and the 4.00pm finish when output remained the same. It would not be unreasonable to expect this to have happened as, by and large, work conditions were being enhanced. Then the researchers removed all the improvements and the women returned to working a 48-hour week, with no breaks or free meals and Saturday working. Output increased further.

At the time this increase was regarded as a mystery and it was only with hindsight it was realised there can be informal organisations within the formal organisation. Further, these informal groups exercise a strong influence over the behaviour of the workers, especially if they are given the freedom to take over responsibility for the way they can work within the group, which is what happened at Hawthorne. Here they were being observed by the researchers rather than being supervised like those not in the selected group. In other words, workers should not be seen in isolation, but as members of a group, in which case supervisors should realise this and see that the group will, if allowed, have common purposes and objectives.

1.5.2 Mary Parker Follett (1868–1933)

Follett was born in Boston, educated at Harvard and Cambridge and studied political science, history and philosophy before becoming a social worker with particular interests in the workplace. She promoted many ideas, many of which were largely ignored at the time, perhaps because they came from a woman at a time when business was very much a male domain. Her profound thoughts and writing are now accepted as relevant today and it is perhaps significant the Japanese hold her in the very highest esteem. She believed organisations would only work well as a whole if all the parts worked together to meet the company objectives. Her views on conflict are interesting and underpin much of her writings. She believed conflict could not be avoided, but should be used to work for us. Indeed, differences could be turned to become a positive asset in an organisation. She advocated that one should not ask who or what is right, but rather to assume both sides are right, viewing the issues from different standpoints having asked different questions. An industrial dispute is an obvious example of this, each side placing a different emphasis or spin on the issue. A dismissed employee will often believe they were unfairly dismissed whereas the employer perceives the 'facts' differently. The resolution of conflict is not to submit to, or have a victory over the other, nor to produce a compromise, but to find and 'integration' of interests or joint problem solving. Her belief was that working with someone was better than working over or under someone.

A summary of these thoughts is given in the following quotation from her writings: 'One person should not give orders to another person, but both should agree to take their orders from the situation. If orders are simply part of the situation, the question of someone giving and receiving does not come up.'

She also postulated that responsible people must be in direct contact irrespective of their position in the organisation. Horizontal communication is equally as important as vertical chains of command, a significant change from the classical school. She went on to say that workers should be involved in policy or decisions when they are being formed and not brought in later. By doing this morale and motivation will be increased. Some 60 years later the construction industry has belatedly adopted this philosophy by involving the contractor in design decisions.

1.5.3 Chester Barnard (1886–1961)

Barnard was President of the New Jersey Bell Telephone Company, he emphasised communication as an important means of achieving goals. He also introduced a new acceptance theory of authority, which argues that subordinates will only accept orders if they understand them, see they are consistent with the aims of the organisation, are in line with their needs and that of their colleagues, and they are willing and able to comply with them. So, for example, if a system of quality assurance is imposed on staff and they are not advised as to the reasons why and its purpose, or they consider it to serve no useful purpose, they will either not use it or sign each stage off without giving the appropriate attention.

He also said that it was no good producing organisation charts if the personnel needed to fill the roles were neither available nor able, and, if necessary, the organisation should be amended to take account of the human resources available.

1.5.4 Abraham Maslow (1908–1970)

A psychologist by background, Mayo eventually became chairman of the psychology department at Brandeis University. Without doubt one of the most commonly referred to theories on motivation is his 'hierarchy of needs'. The basis of his theory is threefold:

- That human beings have needs that are never fully satisfied.
- There are a hierarchy of needs from basic to higher level needs.
- That once one level of need is satisfied the next level needs to be satisfied. However, it should be noted the line between the different levels is blurred. You don't wake up one morning and decide you have been promoted to the next level of need.

He also defined five different levels of need.

1. *Physiological and biological needs*. These are the basic needs for sustaining human life such as food, water, clothing, shelter, sleep and sexual satisfaction. Examples of this are regularly seen in famine areas where those fit enough will fight and scramble to obtain any food available. Once this need has been satisfied they will wait in orderly queues for food to be distributed. This raises them to the second level.
2. *Safety and security needs*. These are the needs that occur once the physiological needs are satisfied, referred to as security, order and stability. People generally like order and predictability in their lives, which is probably why most do not become entrepreneurs. They feel safe, when as a child, their parents provide a safe and secure environment in which to live and as adults being secure in the knowledge that their income is sufficient to pay the bills and provide a reasonable standard of living. This level of need might also explain why religion or other philosophies (such as communism), acting as familiar, well structured and organised environments, tend to me more prominent in the poorer communities of the world.
3. *Social and belonging needs*. This third level need is sometimes referred to as the love need, but should not be confused with sex needs, which are at first level. It is about being accepted by others, feeling part of a group, enjoying and seeking friendship. The need to belong perhaps best sums up this need. In a developed society this is seen very clearly in the development of children when they reach an age when they move from wishing to be solely dependent on their family to developing friendships with other children and even at an early age demonstrating group tendencies in fashion and team games. When no longer permitted in the group as a result of bullying or exclusion, the impact on the child's happiness is clear for all to see.
4. *Esteem needs*. Once people feel accepted they tend to want to be held in esteem by themselves and others. Maslow stressed this should include

esteem for others also. Satisfaction of this need generates such feelings as power, prestige, status and self-confidence. If this need is thwarted, the person enters a feeling of inferiority, lack of confidence and inadequacy and therefore a major lack of motivation. This is important for managers to realise when dealing with their subordinates.

5. *Self-fulfilment and self-actualisation needs.* This is regarded as the highest level as when all the previous needs have been met, people are self-fulfilled and have reached their full potential. At this stage it is likely that people wish to put something back into society also. That is not to say that those at the lower levels would not.

1.5.5 Douglas McGregor (1906 –1964)

McGregor spent most of his working life as professor of industrial management at the Massachusetts Institute of Technology. His most famous concepts are known as Theory X and Theory Y. His view was that managers' assumptions on how workers thought and reacted were at one extreme, X, or at the other, Y, and then treated their workers accordingly. In practice, managers' assumptions are not always as clear as X and Y and fall some way between the two. It is important to stress it is what the managers assume and believe rather than what workers actually are.

- *Theory X.* The average human being has an inherent dislike of work and will avoid it if possible. Because of this human characteristic, most people must be coerced, controlled, directed, or threatened with punishment to get them to put forth adequate effort in order to achieve organisational objectives. The average human being prefers to be directed, wishes to avoid responsibility, has relatively little ambition, and wants security above all.
- *Theory Y.* The expenditure of physical and mental effort in work is as natural as play or rest, and the average human being, under proper conditions learns not only to accept, but also to seek responsibility. People will exercise self-direction and self-control to achieve objectives to which they are committed. The capacity to exercise a relatively high level of imagination, ingenuity and creativity in the solution of organisational problems is widely, not narrowly, distributed in the population, and the intellectual potentialities of the average human being are only partially utilised under the conditions of modern industrial life.

The more enlightened manager favours the Theory Y approach, which, if they take full advantage of, can capitalise on the considerable skill and motivation in the workforce.

1.5.6 Frederick Hertzberg (1923–2000)

Hertzberg was the professor of management in Cape Western University where he established the Department of Industrial Mental Health. During this period he was a consultant to many multi-national companies such as British Petroleum and General Motors. His main study, the motivation-hygiene theory, was recorded in the *Motivation to Work*, its objective to test the hypothesis that man has two sets of needs: first to avoid pain and second to grow psychologically.

- *The Motivation-Hygiene Theory*. Two hundred engineers and accountants, who represented a cross-section of Pittsburgh industry, were interviewed. Each was asked about occasions and the type of experience they had experienced at work, which had either resulted in a noticeable improvement or reduction in their job satisfaction. They were also asked how strongly they felt about each of these experiences, be they good or bad, and the duration they continued to have feelings about them. The areas covered, which generally were satisfiers, were achievement, recognition, the job itself, responsibility and achievement, whereas company policy and administration, supervision, salary, interpersonal relationships and work conditions were measurements of dissatisfaction.

Examples of the results include when people were given responsibility they feel very good about it for a reasonable length of time and whilst not feeling quite as great about achievement, this has a very long lasting effect. On the other hand, all those administrative rules and regulations cause irritation for some considerable time and lack of recognition gives a strong feeling of dissatisfaction, but over a shorter period of time.

1.5.7 Renis Likert (1903–1981)

Renis Likert was an American social psychologist. He was the first director of the Institute for Social Research at the University of Michigan and is acknowledged as one of the leaders of research into human behaviour in industrial organisations. He wrote on many subjects, but of particular interest are his findings on the way the style of supervision can affect productivity, and his four definitions of management style.

He investigated how employee-centred and job-centred supervisors have an impact on performance. To measure this he used several criteria to evaluate their administrative effectiveness. They included productivity per man-hour, job satisfaction, turnover and absenteeism, costs, scrap loss, and employee and management motivation. He clearly established that supervisors who concentrate on the job itself, by breaking down the job into simple clearly defined tasks, determining the method of work, employing and training personnel to do the specific task and check they are performing to the standards set, generally are found to have a lower output rate than those supervisors who are interested in the personnel. In this case they are more interested in their problems and building effective working teams. This means giving the operatives the opportunity to make decisions for themselves, delegating responsibility and leaving them to get on with the job rather than continually monitoring performance.

The four management systems he identified, which can be found in organisations are:

1. *Exploitative–Authoritative.* This is when top management is very autocratic, makes all the decisions, motivates using threats and discourages any input from below. Senior management takes most of the responsibility, the lower levels have very little. What communications exist are top-down resulting in little teamwork. Subordinates do not feel free to discuss anything about their work with their manager. This is a management style based on the premise stated in McGregor's X Theory (section 1.5.5).
2. *Benevolent–Authoritative.* As with the exploitive style, senior management makes most of the decisions, although in this case some lesser decisions may be made at lower levels. A condescending attitude is usually displayed in communicating with subordinates, resulting in a subservient attitude towards superiors. There may be some minimal flow of ideas from subordinates to managers, but generally there is very little communication between the two. Many of the old family firms adopted this benevolent attitude where members of the family would be referred to by their Christian name always prefixed by Mr. They often felt they had a responsibility for their employees.
3. *Consultative.* Although senior management still reserves the right to make decisions and control the business, ideas are sought from below. Usually they only have a certain level of trust in their subordinates and are therefore reluctant to fully delegate to them. As a result there is some two-way communication and some teamwork. Whilst subordinates feel relatively free to discuss things about their job they can become frustrated as often their opinion is ignored if it does not coincide with that of the manager. In many ways this is the worst management style to be a subordinate too.

4. *Participative.* This is by far the most satisfactory work environment. In this case senior management has complete confidence in its subordinates and is prepared to fully delegate responsibility to them. This results in higher levels of satisfaction, although sometimes more sleepless nights as a result of the responsibility. Subordinates feel completely free to discuss issues with their managers and equally managers ask for advice and opinion. This develops a strong group and teamwork spirit. In the ultimate case managers will accept a strong majority view even though it is not the same as their own. There will of course be occasions when the manger feels so strongly that he or she will overrule the group, but experience indicates the rest of the group will accept this knowing in the end the manager has to take the responsibility. Generally, Likert found that managers adopting this style were the most successful.

1.5.8 Robert Blake and Jane Mouton

Both American psychologists, Blake and Mouton were president and vice-president, respectively, of Scientific Methods Incorporated, providing behavioural consultancy services to industry. They developed their Managerial Grid, which, similar to Likert's, combines two fundamental aspects of managerial behaviour, that of concern for people and concern for production. The term production is meant to cover all tasks, be it manufacturing, construction, volume of sales or number of accounts processed. The grid demonstrates there are a variety of combinations for the degree of concern for people and production. These are:

'Country Club management' has a high degree of concern for people, but little concern for production. On the positive side, as with participative style of management (section 1.5.7), personnel are encouraged and supported by management, mistakes are overlooked on the grounds that the person making them is aware of this, and there is an overall camaraderie which helps the production process. On the downside there is the danger that people avoid conflict, and problems are watered down rather than being properly addressed. Any new ideas someone has that might be controversial tend not to be brought forward for discussion as this might cause upset.

'Impoverished management', where there is neither concern for people nor production, is a ticket to organisational failure. The supervisor avoids responsibility and blames his superior for any difficult or unpopular directive or tells the supervisor it is the personnel below him who have made the mistake. On the surface it seems hard to believe this kind of management could exist, but people that have been passed over for promotion can show this tendency either for a short or prolonged period. This is often referred to as 'taking one's bat home'.

'Task management' focuses primarily on production and has little concern for people; defined by Likert as job-centred. This type of manager expects people to do as they are told without question and for programmes to be met on time. In the event of something going wrong they will look for someone to blame. They will not accept any disagreement and it will be perceived as insubordination. It is suggested this style of management was reflected in the struggles between the trade unions and certain companies especially in the 1960s and 1970s, which resulted in strengthening unions and greater industrial unrest.

'Team management' is focused on concern for people and production and believes these two requirements are compatible. Everybody in the team can contribute in some way to achieving the ultimate goal of higher achievable levels of productivity and in doing so satisfies their own needs. The manager's role is to ensure work is planned or organised using the expertise of the others in the team resulting in their full engagement in the process and goals. When conflict does occur, and it almost certainly will, then it should be confronted openly and not seen as battle of sides, but rather as a mutual problem that can be satisfactorily be resolved (section 1.5.2). Generally, this will be the most effective style.

'Dampened pendulum' represents what often happens in practice: that managers swing between 'country club' and 'task management' styles. They start by being much focused on production and when the levels of discontent and arguments arise, move to concentrating on the concern of people at the expense of production and as this improves move back towards production focus and so on. This produces a satisfactory level of production and generally keeps morale at an acceptable level, but does not reach the full potential available.

1.5.9 Victor H. Vroom (1932–)

Victor Vroom was born in Montreal, Canada. He is Professor of Psychology at Yale School of Management and is a leading authority on the psychological analysis of behaviour in organisations. Among many significant contributions to management thought, he is particularly noted for his Expectancy Theory of Motivation in which he examines why people choose to pursue a particular course of action. He argues there are three main issues people consider before expending sufficient effort to complete the task at the appropriate standard. Figure 1.3 demonstrates the basic components and their interrelationship.

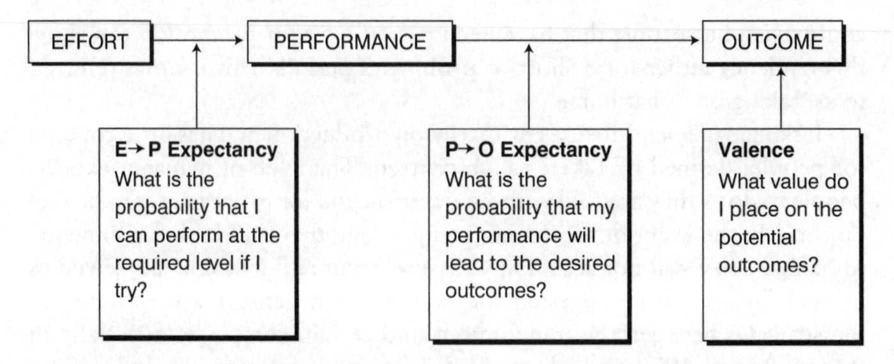

Figure 1.3 Basic components of expectancy theory (adapted from Bartol and Martin (1994) © McGraw-Hill Companies, Inc.)

Effort→performance (E→P) expectancy is a self-assessment of the probability that the efforts put in will realise the required performance level. The assessment of probability will depend on the person's own belief in their ability to accomplish the task and the availability of resources of any kind it is believed are necessary to carry out the work. As an example, a newly promoted project manager may feel ill-equipped to run a type of project they have had no experience of, whereas if the project is similar to others they have worked on before promotion they will feel much more confident in completing the task satisfactorily. In the first case the E→P expectancy is low and in the latter high. This is only the first stage of the process.

The second stage is known as performance–outcome (P→O) expectancy. In this case, it is an assessment of the probability that successful performance will lead to certain outcomes. On the positive side these could include enhanced prospects of promotion and extra remuneration, known as extrinsic rewards, and job satisfaction, recognition and self-development, known as intrinsic rewards. Against this there is the consideration that perhaps having to work longer hours will affect family life and increase pressure and stress. The likelihood of these desired outcomes are then assessed against previous experience within the organisation, such as if hard work is acknowledged and rewarded. If this always happens the P→O will be high, if never, then low, and if sometimes, then the assessment may only be 50/50 or some other proportion.

The third part of the process is called valence. This is an assessment of the worth or value of the anticipated outcome. As indicated before there can be positive outcomes such as extra money, and negative such as loss of time at home. The strength of feeling to these will determine whether the valance is high or low. If it is very positive then motivation will be high. This is a personal judgement. A struggling student will more likely be motivated to take a mundane job in their spare time to make ends meet than if their parents send money regularly. A rich person may carry out volunteer work because of the satisfaction obtained as a result of helping others.

The hypothesis is that people will only put effort into a task after considering these three elements: E→P expectancy, P→O expectancy and Valence. This is called the expectancy theory:

$$(E \rightarrow P) \times (P \rightarrow O) \times \text{Valence} = \text{Motivation}$$

It can be seen from this formula that if any of the three parts of the process is zero or approaching it, then the motivation will also be zero or close to it. Equally, if they are all high then motivation will be high. The outcome of the formula is not the same for each person and management should be aware of this. Each individual has different perceptions of his own ability and the performance outcomes will vary dependent upon personal circumstances and aspirations. This means that personal knowledge of the employee will help considerably especially when evolving personal development plans and motivation.

1.6 Systems theory

Up until the 1950s there had been relative stability in society in spite of the Second World War, but this was to dramatically alter in the 1960s. There was a major revolution in every walk of life as the developed world came out of restrictions placed on it by rationing and so on. In the arts came The Beatles, The Rolling Stones, Elvis Presley, 'That Was The Week That Was', 'Beyond the Fringe', Mary Quant, David Bailey, Harold Pinter and so on. Students demonstrated by sitting in at British universities, they took to the streets in Paris and the Campaign for Nuclear Disarmament marches to Aldermaston started. Trade unions showed their muscle, notably in the car industry, the press and docks, and management was often unable to react positively and effectively with the resulting industrial unrest.

A new management approach was needed, now called the systems approach. The main developers of the systems approach were Richard Johnson, Fremont Kast and James Rosenzweig. They defined the definition of a system as 'an organised or complex whole: an assemblage or combination of things, or parts forming a complex or unitary whole'. Where previous management thinkers looked at individuals, groups or components of the organisation, the systems analysts look at the organisation as a whole and view it as a system with the different parts interacting with each other. An analogy is with a healthy human body in which each organ is related to,

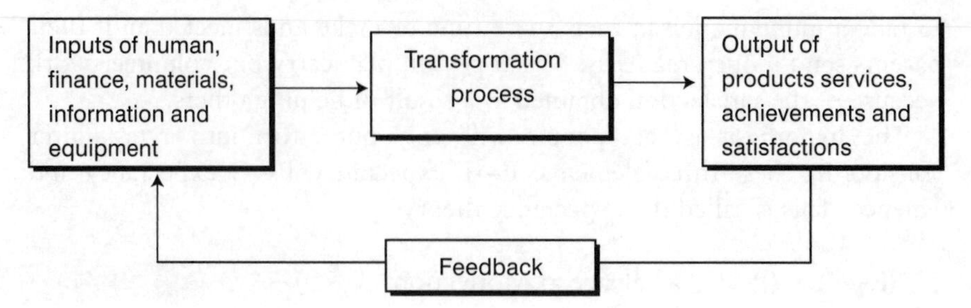

Figure 1.4 The systems approach (adapted from Bartol and Martin (1994) © The McGraw-Hill Companies, Inc.)

interacts with, and is dependent on the proper functioning of all the other organs. If any of these organs malfunctions then the result can, in extreme situations, be total body failure.

A system is made up of four components as shown in Figure 1.4. 'The inputs' are from various sources such as human, financial, materials, equipment and information. 'The transformation process' is the operations and processes where the inputs are converted into, and are a function of, the management and technical abilities of the organisation. 'The outputs' are products and services, but also include the profit or loss achieved and the levels of satisfaction employees obtain. 'Feedback' is an assessment of the results achieved, the organisation's reputation in society and reaction from society, and the working environment both externally and internally. The external environment includes customers, shareholders, trade unions, government and the general public.

There are two types of system: closed and open. A closed system is one that operates within itself and has little or no contact with the environment around it resulting in little or no feedback. The only truly closed system is the universe. An example of the nearest to a closed system in practice is a domestic hot water heating system, but even this has some interaction with the outside when radiators need to be bled or maintained. It would be very difficult to imagine a closed system for an organisation as most, if not all, must have some communication with the outside otherwise what is the point of their existence.

An open system usually has a dynamic and continual interaction with the external environment and between the component parts within the organisation. Whilst the inputs and outputs are very important, the crucial element is the transformation process where there are large numbers of subsystems to be considered and analysed. The system must be adaptable to meet the demands of the external environment by continuously amending its internal systems. This may mean modifying the product or service, reacting to competition, market demand and price. These in turn may affect the inputs such as quality of materials, equipment used and amount of labour. This may require training in new techniques and staff development in general, adapting total quality assurance systems (*Operations Management for Construction,* Chapter 8) and dealing with the management of change (Chapter 7).

There are three main objectives of the systems approach to management. The first is to define the relationships between the various parts of the organisation with each other and with the outside environment. The second is to establish how these relationships work, and the third to establish the purpose of these relationships.

Kast and Rosenzweig postulated there were three main characteristics of open systems, which they called negative entropy, differentiation and synergy. Entropy is the rate of decay measured over time, so negative energy, in an open system is the ability to bring new energy in to the process to slow down or stop the rate of decline. Differentiation is when the system becomes more complex as a result of adding new units to the organisation to cope with the changes and new challenges, for example to satisfy government requirements, employment policy, quality assurance, safety and environmental issues. Note there is a danger that these can increase to such an extent that the administration becomes top heavy so they must always be kept under review. Synergy is the concept that the whole works better than the parts, so if the parts are working and interacting properly the business will perform better.

1.7 Contingency theory

Systems approaches, however, are not the total answer either, which is not surprising with the rapid changes in our society. Modern management thought is moving to the contingency approach; combining the best aspects of all the other theories that have gone before including those which preceded the classical approach, that of trial and error. In other words, the need to be adaptable and flexible and to continue to seek new ways to manage the organisation.

The overriding principle is that the managerial action taken is determined by the particular circumstances of the situation rather than by using universal principles that apply to every situation. It is worthwhile noting that Parker Follett, had suggested this years before, but had been largely ignored.

1.8 Further thoughts

To bring this section up to date it is important to mention the concepts of lean thinking, benchmarking, best practice and Theory Z. Much work and research in all of these areas has been done. It is not the intention to develop these ideas further other than to say that lean thinking is primarily to do with the banishing of waste of all types, not just material waste. Benchmarking is concerned with comparisons between organisations and establishing key performance indicators (not just necessarily in one sector) and best practice, linked to benchmarking is concerned with establishing the best way to carry out or organise and control work.

William Ouchi has suggested a way to combine the Japanese way of running organisations with that of the American to get the best of both worlds. He called this Theory Z. In the future the author suggests the impact of sustainability issues will become a further stage in the development of management thinking. Whilst nothing written before will become obsolete, personal and society values will change and there will be a move towards much more self-reliance and regional sustainability. This may well be in conflict with the shift in economic power from the current developed world to that of China and India.

References

Bartol, K.B. and Martin, D.C. (1994) *Management*, 2nd edn. McGraw-Hill.

Blake, R.R. and Mouton, J.S. (1985) *The Managerial Grid III: The Key to Leadership Excellence*. Gulf Publishing.

Fayol, H. (1949) *General Industrial Management*. Pitman & Sons.

Graham, P. (ed.) (1995) *Mary Parker Follett: Prophet of Management: A Celebration of Writings from the 1920s.* Harvard Business School Press.

Hertzberg, F. (1959) *The Motivation to Work*, 2nd edn. John Wiley.

Johnson, R.A., Kast, F.E. and Rosenzweig, J.E. (1963) *The Theory and Management of Systems*. McGraw-Hill.

Kast, F. and Rosenzweig, J.E. (1972) General systems theory: applications for organisation and management. *Academy of Management Journal*, 15(4): 447–65.

Likert, R. (1961) *New Patterns of Management*. McGraw-Hill.

Luthans, F. (1976) *Introduction to Management: A Contingency Approach*. McGraw-Hill.

Maslow, A. (1954) *Motivation and Personality*, Harper & Row.

Mayo, E. (1933) *The Human Problems of an Industrial Civilization*, Macmillan.

McGregor, D. (1960) *The Human Side of Enterprise*, McGraw-Hill.

Megginson, L.C., Mosley, D.C. and Peitri, P.H. (1989) *Management Concepts and Applications*, 3rd edn. Harper Row.

Ouchi, W.G. and Jaeger, A.M. (1987) Theory Z organizations: stability in the midst of mobility, *Academy of Management Review,* 3(2): 305–14.
Peters, T.J. and Waterman, R.H. (1982) *In Search of Excellence: Lessons from America's Best-Run Companies*. Harper & Row.
Taylor. F.W. (1947) *Scientific Management*, Harper & Row.
Vroom, V.H. (1995) *Work and Motivation*, Rev edn. Jossey-Bass Classics.
Weber, M. (1947) *The Theory of Social and Economic Organization*. Free Press.
Womack, J.P., Jones, D.T. and Roos, D. (1997) *The Machine that Changed the World*, Macmillan.

CHAPTER

2 Organisations

2.1 Introduction

The purpose of this chapter is to explore some of the basic concepts of organisations and how they come to exist. Organisations are formed from groups of individuals who have different authority and responsibilities which, when working well, allow the business to meet its objectives. The authority and responsibility given to people is the basis for managing the organisation. Some of these relationships are formal and others informal.

Organisations need to group individuals into particular functions with clearly defined tasks with someone in charge – the manager or supervisor. How authoritarian this managerial role appears is dependent upon the company's management style or the type of business. The armed services will have less debate within the group about decisions than a creative advertising team. These groups may also be subdivided further. For example, the overall group may be 'the production activity' but in construction these would be divided into each individual site, and in each site there could be further divisions of say, each block of flats.

The organisation must take account of how communications are best effected up and down the business, but also needs to consider how horizontal communication can take place. In construction, these horizontal lines of communication are very important and often difficult to enact. The very nature of construction is that each project is different and many of the participants in the process may never have worked together before. Whilst the project manager may select certain key staff, the regional or head office senior quantity surveyor and engineer will have selected their representatives. To make matters worse, the majority of the construction work is carried out by sub-contractors, whose primary concern is to complete their particular piece of the work. Their prime line of communication is directly to the main contractor. On top of this, the design team can also be appointed separately without any consultation with the contractor, although some of the new forms of contract procurement such as management contracting and design and build, have reduced the problem.

2.2 Span of control

Before designing an organisation structure it is necessary to consider for how many people an individual can be responsible. Generally, as organisations expand, more employees are employed to carry out the business. Whereas in a very small business the owner makes all decisions, in a larger organisation many of these decisions have to be delegated. Thus the manager of a group is given certain responsibilities and authority. The reason for this delegation of power is simply because the owner begins to lose control if too many people are responsible to them. In other words there is a limit to the number of subordinates a manager can control. This is known as 'span of control'.

So how many people can one manger control? In theory spans can vary from one to many tens of subordinates. However, it is generally accepted as a rule that the most likely range is between 3 and 20. Some people argue the range should only be from 5 to 7 to make the organisation more efficient, but to be too prescriptive is dangerous. What is significant is that the size of span affects the style of the organisation. If the span is large the lines of communication between the higher and lower levels in the organisation are short, and if the spans are small, the lines are longer. These are known as tall and flat structures (section 2.6).

There are various factors that can influence the size of the spans of control, such as:

- Senior management may take the decision and decide the numbers because they want a particular type and shape of organisation.
- The talent, personality and ability of the manager to cope under pressure can be a determining factor.
- The experience of the manager is very important. If a manager has been in post since the process started, their accumulated knowledge and responsibility can be extremely high and it is often only realised how much when that person is being replaced. The replacement can be by more than one appointment.
- If the ability and knowledge of the subordinates is high they will require less supervision thus the manager can control more; a less able work force would require more supervision.
- If the work is very complex and there are lots of changes to the work occurring regularly, increased supervision will be required.

- Equally, if quality levels have to be high, extra supervision may be required especially if the workforce is not of the highest calibre.
- If the subordinates are located a long way from base and it takes the supervisor a long time to reach them, the span of control will reduce. This is not unusual in construction, where a sub-contractor's workforce may be spread over several contracts in small gangs, each not requiring full-time supervision.
- Closer supervision is required if the task has a higher than normal level of risk.

2.3 Organisation charts

Organisation charts are diagrams showing the inter-relationship between individuals or groups of people. They can be used to demonstrate how the organisation groups its various activities, and the relationships between functions and tasks. The chart demonstrates who is responsible to whom within the organisation known as the chain of command. It can also show the horizontal links between functions and groups. The horizontal lines are very important and do not necessarily mean that because two people are on the same level they are of the same seniority.

The author has observed that organisation charts tend to be produced in two different ways. The first commences at the top with senior management deciding who should be responsible to whom until the bottom of the chart is reached. The other, and the author believes is the more successful, starts at the bottom of the organisation and asks the question: 'What does this person or group need to best complete their task?' This works on the assumption that in manufacturing and construction as the operatives are the producers and it is their level of output that determines the viability of the business, their needs have to be carefully considered (section 2.7).

2.4 Types of organisation

There are many different ways organisation charts can be compiled; this section describes the most common.

2.4.1. Executive or line responsibility

The military is probably the closest to this form with the chief of staff at the top, with the chain of command descending through the ranks to the private, in the case of the army. In construction the executive function will exist typically as shown in Figure 2.1.

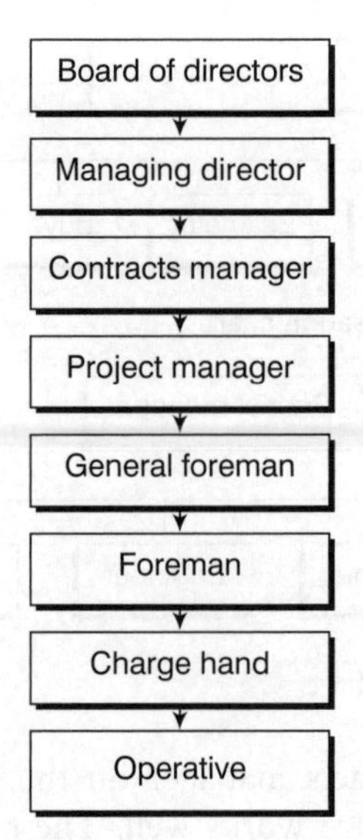

Figure 2.1 The executive function

This assumes the company employs all the people involved in the process. In the UK this would be a rarity, so the chain of command might stop at the general foreman. However, sub-contractors often do employ all of these titles, so that chain of command would be similar, although the smaller businesses would have fewer levels of management. In this example the project manager is the person responsible for management of a construction site and the contracts manager is the senior manager responsible for all current construction projects. This executive line will general exist in all businesses, but cannot exist in isolation. The people in this chain of command need support to carry out their tasks.

2.4.2 Functional responsibility

It is often stated 'you should only have one master', and on the surface this makes sense, but on further investigation it is found this does not always work. In construction it is quite common for a person to have two supervisors, but each of their responsibilities for the subordinate is clearly defined.

A typical head office organisation chart might look like Figure 2.2, with the senior quantity surveyor, planner, estimator, buyer, contracts manager and so on reporting to the managing director. On site the organisation chart might be as shown in Figure 2.3.

Take for example the quantity surveyor (QS) on the site who can have a responsibility to two different people, the chief quantity surveyor in the

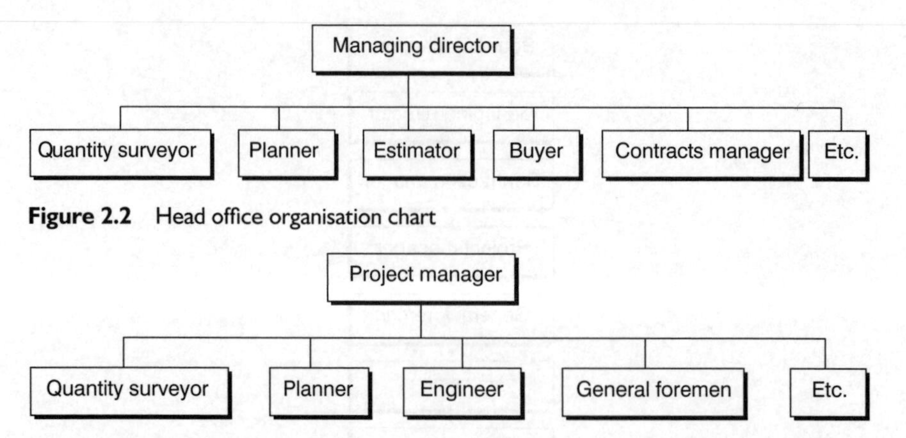

Figure 2.2 Head office organisation chart

Figure 2.3 Site organisation chart

head office and the contracts manager on the site. This may be seen as contradiction, but in practice works well. The project manager has to run the project and is responsible for the general behaviour of the staff, such as punctuality, appearance and attitude, but may have limited knowledge about the professional competence of the QS. The monthly valuation may be completed on time, but was it done properly? The manager can ask for information by a certain time, but how reasonable is the request? The responsibility of the chief QS is to ensure that the site QS is providing a proper and reliable service.

What happens if the site QS takes issue with the project manager and it cannot be resolved at this level? The project manager or QS would normally go to the chief QS and try to resolve the dispute. If this did not result in a solution, the project manager could go to the contracts manager at the head office who would then hope to resolve the issue with the chief QS. In the unlikely event of them being unable to sort it out they would both have the option of going to the managing director for resolution. This identifies an important issue in the construction of an organisation chart in that it can demonstrates the lines of communication in the event of arbitration between two parties being required.

2.4.3 Product structure

Large organisations often divide their structure into product divisions. Schools have science, English and language departments, hospitals divide their work into outpatients, accident and emergency, ear, nose and throat and so on. Construction companies used to divide their business between

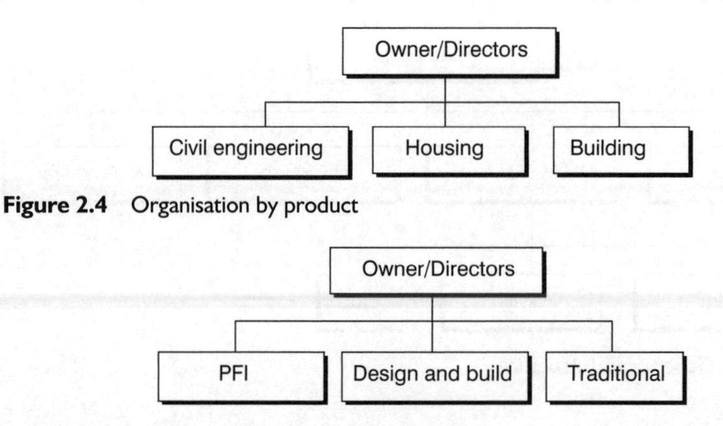

Figure 2.4 Organisation by product

Figure 2.5 Organisation by procurement method

types of work such as civil engineering, housing and building projects, as shown in Figure 2.4. But in recent years they have tended to divide between procurement methods such as Private Finance Initiative (PFI), design and build, and traditional, as in Figure 2.5, each with their own subdivision reflecting their specific needs.

Although not shown on these charts there will be certain activities probably carried out centrally on behalf of the group, such as human relations management, purchasing, public relations and marketing.

The advantage of adopting a product approach is that attention is focused on each category, which might have different needs and expertise from the others. For example, housing is in a speculative market anticipating customer requirements, where traditional contracting work is obtained by competitive tendering and then constructing what the client has already determined. The contractual differences between the various methods of procurement are significant and PFI work may also include the facilities management element. The disadvantage is that there can be duplication of work. This is why certain functions will be carried out centrally.

2.4.4 Geographical structure

Geographical structures are produced, especially in large organisation, based on where the company's activities are. If a construction company is working in most parts of the UK there becomes a problem of communications if everything is controlled from one location. Local knowledge is important for many reasons. Senior management builds up a network of local connections and thus potential clients. The local market is better understood, as are the availability of competitive and reliable sub-contractors and material

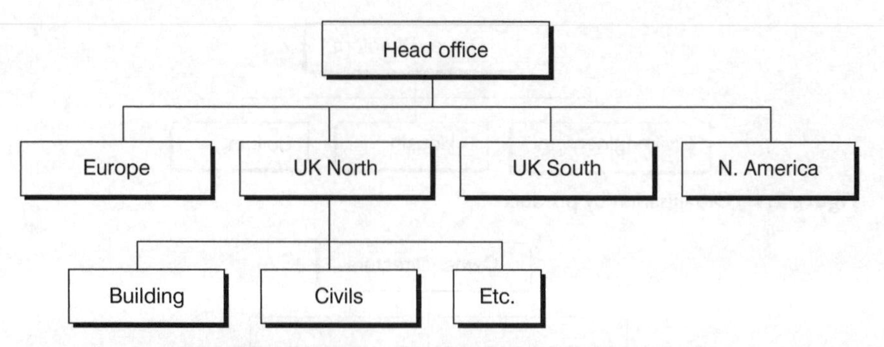

Figure 2.6 Geographical structure

suppliers. The same applies internationally, but there are also the issues of time differences, different regulations and laws, and understanding local culture. As with product structures, certain activities will be carried out centrally. Figure 2.6 demonstrates a typical example.

2.5 Centralisation and decentralisation

Centralisation is defined as concentrating the power and authority near or at the top of the organisation; decentralisation is delegating the power and the authority to make decisions to lower levels of the organisation. No organisation is entirely one or the other, unless it is a one-man business, but will tend to be primarily one or the other.

Which type an organisation is will depend on a variety of factors:

Management's philosophy

Some managers believe in strong central control and want to be in total control. They build up a close team around them of quality people and make all the major decisions. The lower levels in the organisation are then instructed what to do as with Likert's exploitive and benevolent management styles. Many of these managers also believe the staff below fall into the category of McGregor's Theory X. Other managers tend more towards Likert's participative management style and believe employees fall into the category as described in McGregor's Theory Y. They delegate much of the decision making and accompanying responsibility and accountability to those who have the appropriate information available to make the best decision.

Organisational growth

Organisations that grow and remain centralised do so usually as a result of the way the company was set up in the first place. When organisations merge or expand as a result of further acquisitions they tend to move towards being decentralised. This is for a variety of reasons which include: different cultures of the businesses; different products or services; the fact management structures are already in place, although these are likely to be radically changed as part of the merger and takeover process; the businesses will be geographically spread; and the sheer scale of the business.

Geographical diversity

The greater the geographical diversity of the business the more likely the organisation is to be decentralised. This is due to the problems of control from distance. For example, a construction company working nationally will set up regional organisations staffed by personnel who know the local market and supply chain and can therefore make informed decisions. International organisations have an added problem of the time differences between the different parts of the organisation, not to mention the customs and practices of another country. In construction, regional offices may be given a limit to the price of a tender because of the risks involved, and contracts over a certain size may be considered centrally.

Effective controls

Organisations which have difficulty in controlling the lower levels, tend to be centralised. This can occur within an organisation which is primarily decentralised, especially if the work occurs at random, such as with a maintenance department of a factory, where there is a need to move personnel rapidly from one job to another at a moment's notice depending on the urgency of the breakdown. Decentralised organisations means that if you delegate the work there must be some control to ensure what has been delegated is being carried out properly.

Quantity and quality of managers

By definition, if the business is decentralised and powers delegated, there have to be sufficient qualified and competent managers to take responsibility and to make sound decisions. Universities are examples of where there can be a limited number of good and qualified managers since traditionally promotion on the academic side is primarily a function of scholastic ability rather than management ability. This creates a problem for senior management: should they have many faculties, with a few good managers running them with lots of small departments run by less able managers, or should there be fewer faculties and departments, to better use the limited management resource.

Diversity of products and services

The contracting industry is diverse and is changing continually. Traditionally, there was the split between civil engineering and building contracting, but over the years the line between the two has become blurred, and qualified civil engineers and builders work on both types of processes. Speculative housing has still remained as a separate unit. The traditional form of procurement has given way to others, such as design and build, management contracting and Private Finance Initiatives, each requiring different skills and knowledge. This means there is a tendency for national companies to decentralise, up to a point, to cope with these diversities.

It is interesting to note that many organisations go through cycles of moving from centralised to decentralised and vice versa approximately every two decades. The author is reminded of the following quotation: 'There are those who believe that the best way to meet any new situation is by reorganisation. This is a wonderful method of creating the illusion of progress while producing confusion, inefficiency and demoralisation.' (Petronius Arbiter, AD 65)

2.6 Tall and flat structures

A flat structured organisation has fewer levels in the hierarchy, but with many more subordinates reporting to one manager with a wide span of control. The tall structure has many more levels in the hierarchy, but with fewer people reporting to a manager and with a narrow span of control, as shown in Figures 2.7 and 2.8. In each example the same number of boxes have been used in the diagram yet the flat structure has only three levels of management whereas the tall structure has five.

In the tall structure the flow of information vertically is interrupted many times more than in the flat structure, but in the latter there tends to be more discussion and consultation. So there is a balance between, on the one hand, more time for the flow of information to take place, as against more time required to both discuss issues and co-ordinate many subordinates. The taller structures tend to be more dictatorial than the flatter structures as a result. However, because in the taller structures the span of control is small the

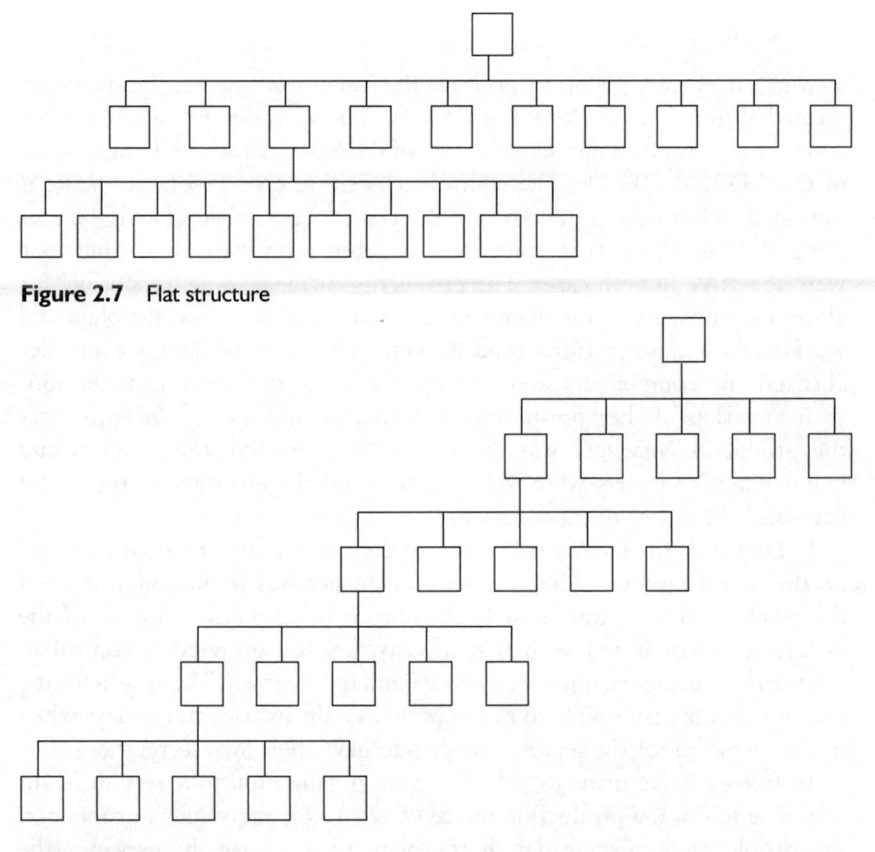

Figure 2.7 Flat structure

Figure 2.8 Tall structure

manager has more time to consider issues as less time is spent supervising. The lower levels in the hierarchy in the flatter structures usually know the senior managers much better than in the taller structures, which can increase the quality of personal contact. Taller structures are more centralised organisations than flat structures. To manage a flat structure, because of the size of the span of the control, there needs to be more delegation of responsibility and this tends to result in higher morale of the employees. It is a complex issue to determine which of the two is the better. Although there is some evidence the taller structures perform better on measures of profit and return on investment (Carzo and Yanouzas 1969).

2.7 Case study

Two factories were set up to produce the same product, but the managers created different organisation structures. The factories produced precast concrete structural components for one of the industrialised building systems of the 1960s and 1970s. The equivalent position on a construction site is indicated in brackets in the boxes in the chart where appropriate in Figures 2.9 and 2.10. There were other roles not shown on these charts, but they were the same in both cases. The production manager was responsible for all production activity, the maintenance engineer ensured that the plant was working to as close to 100% capacity as possible, and the quality controller checked all components were being made to the correct dimensions with materials of the appropriate performance and quality. In both cases the production manager was the second in command, the planners and maintenance engineers were well qualified and the quality controllers had previously been one of the foremen.

In factory 1, the factory manager had determined the organisation based on the following logic. The production manager was responsible for all of the production so it was sensible for him to be responsible for all of the departments connected with that activity, i.e. the production controller, maintenance engineer, quality controller and the foremen. The reason for the secretary being responsible to him is perhaps a throw back to the days when it was the norm for the senior managers to have their own secretaries.

In factory 2 the manager saw the same organisation in a very different way. The job of the production manager was to obtain as much production as possible each working day. If the plant broke down, he expected the maintenance engineer to make a rapid temporary repair and if necessary

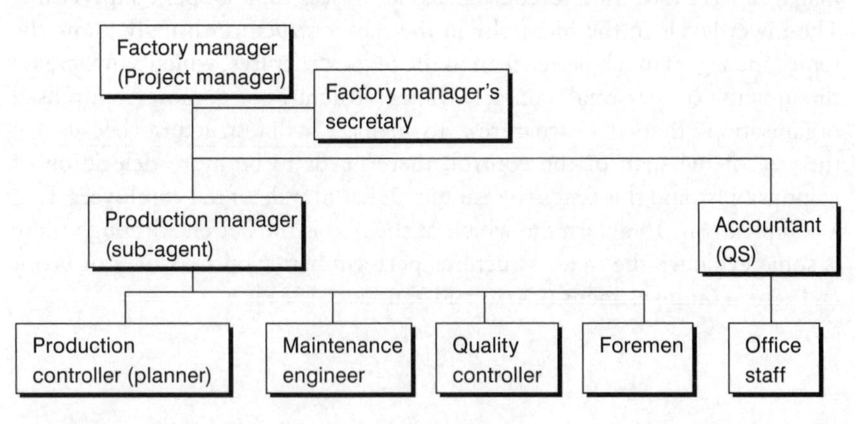

Figure 2.9 Factory I

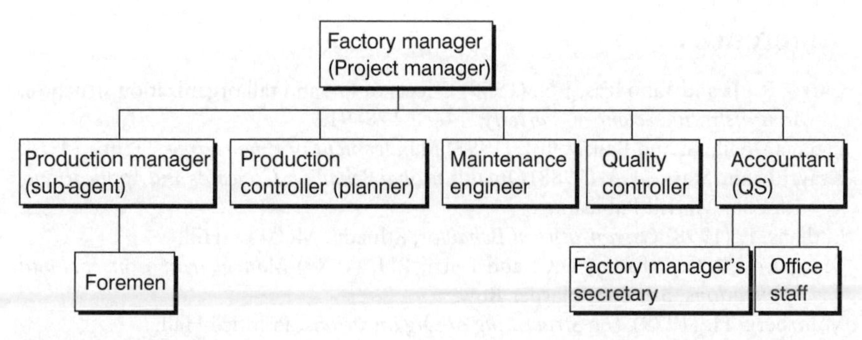

Figure 2.10 Factory 2

work at night to make the repair more substantial and lasting. The engineer, on the other hand, would want to make the repair properly at the time of the breakdown. The quality controller acted as a brake on the production manager. As production rates increased, quality standards would deteriorate. The production manager leaned towards being prepared to make sub-standard components, if it kept productivity moving and if they could be rectified afterwards. The production controller had to ensure there were sufficient units of the type required by the site to be ready in store for when the site called for them. Ideally, the production process would prefer to make all of one type then change to the next, because every change would stop production whilst the changes to the plant were being made. The operatives then would have to go through a learning process that could diminish production outputs. All this meant that there was potential conflict between the production manager and the others.

By putting all of these roles on the same level, it was incumbent on each to resolve their differences. If they were unable to they could go to the factory manager for arbitration. Most people would prefer to solve their problems rather than go to their superior for action. In factory 1, the production manager could tell the others what to do and whilst there would be some discussion, in the end he made the decision, which did not ensure that the best solution was always used.

In factory 2 the factory manager soon realised that he did not produce enough work to keep his secretary fully occupied. He decided to make her responsible to the accountant, who was also the office manager, her unused time could then be put to good use. He would still have priority call on her when he required work to be done.

References

Carzo, R., Jr and Yanouzas, J.N. (1969) Effects of flat and tall organization structure. *Administrative Science Quarterly*, 14(2): 178–91.

Freeman-Bell, G. and Balkwell, J. (1993) *Management in Engineering*. Prentice-Hall.

Gray, J.L. and Starke, F.A. (1988) *Organizational Behavior: Concepts and Applications*, 4th edn. Merrill Publishing.

Luthans, F. (1998) *Organizational Behavior*, 8th edn. McGraw-Hill.

Megginson, L.C., Mosley, D.C. and Peitri, P.H. (1989) *Management Concepts and Applications*, 3rd edn. Harper Row.

Mintzberg, H. (1979) *The Structuring of Organizations*. Prentice-Hall.

CHAPTER

The legal establishment of businesses

3.1 Introduction

There are a very large number of organisations involved in the construction industry. They range from multinational companies or corporations, such as Bovis Lend Lease and Skanska, down to one-man businesses that carry out minor repairs to property. There are partnerships, especially in the design sector, varying from a couple to those with several tens of principals. They are all in business to make a profit, but how they are constituted legally differs considerably. Whichever type of construction business it operates in, it is inevitable that contact will be made with other businesses with different legal constitutions. The majority of businesses are sole traders, partnerships and corporations, but this section will also look at co-operatives and franchised businesses.

3.2 Sole traders

This is the simplest form of business and therefore the most common. It is the establishment of a business with a self-employed single owner (sole trader). Examples include small shops, plumbers, electricians and consultants. It is simple to set up, just start trading, but sole traders have to register with HM Revenue & Customs (HMRC) for which there is no charge. There is a simple form to complete and submit to the HMRC.

They can use their own name or another providing it is not the same as another company's or one that could be interpreted as such. There was the case when a garage owner called his business MGM, after his initials, but the

American company of the same name took exception to this and took him to court to change the name of his business. The sole trader has no special legal status, as there are no particular laws other than those that apply to any citizen.

The key features are:

- The business annual accounts which have to be produced for tax purposes, do not have to be disclosed publicly and are therefore not available to the competition.
- All profits are taxed as income. The fixed-rate Class 2 National Insurance contributions (NIC) must be paid regularly and Class 4 NIC on any profit.
- The owner makes all the decisions on the management of the business and therefore has total control. The disadvantage is the quality of the decisions is a function of the competence of the owner, as they may have no one to discuss it with.
- They can offer a personal service to customers.
- All profits and debts belong to the owner; the former can be spent without any justification to anybody. It is more difficult to raise funds/loans, and bank charges and interest rates can be quite high as they argue the need to do this because of the higher risk. The family home may be the only significant asset and if used as collateral may have to be sold to service debts, along with other assets.
- The owners have to carry out all the administration themselves, which may entail long hours.
- Sickness and leave, for the owner, and when children are on school holiday or ill, can often cause problems without cover available.
- Often it is not possible to purchase products and services cheaply because of the economy of scale, whereas a large contractor has considerable purchasing power.

3.3 Partnerships

If two or more people wish set up a profit-making business they can form a partnership. They become part owners of the business and share the responsibility of managing the business, share in its profits and are liable for any losses. In construction such businesses commonly encompass quantity surveying, building surveying, architects and engineers. There are two types of partnerships known as ordinary partnerships and limited liability partnerships.

3.3.1 Ordinary partnerships

There are three main types of partner called general partners, sleeping partners and companies. General partners invest their money into the partnership, actively assist in running the business and, for their efforts, take their share of the profits. Equally they are fully liable for any debts the business may accumulate, even if these are greater than their initial investment, so their personal assets could also be at risk. They must register themselves as self-employed with HMRC. Any partnership must have at least one general partner. Sleeping partners invest money into the business, but take no part in running the business, but they share in the profits and are also fully liable for the partnership's debts. Companies can become members of a partnership and have the same rights and responsibilities as other types of partners. Each partner is jointly liable for any contract made by any other of the partners, irrespective as to whether or not they had prior knowledge.

It is usual for partnerships to have an agreement or a 'deed of partnership', which is a legally binding agreement between all the partners describing how the business will be run. It will normally include who has which responsibilities, whether a partner is to receive a salary, the time they are to spend working, what happens if a partner dies or wishes to leave the partnership, how new partners can be appointed, and the apportion of any profits. A partner, who for example, has put in the greatest amount of capital initially, might expect to take a greater share of the profits.

The main features are:

- The business annual accounts which have to be produced for tax purposes, do not have to be disclosed publicly and are therefore not available to the competition.
- As the partners are self-employed all profits received are taxed as income, the fixed-rate Class 2 NIC must be paid regularly and Class 4 NIC on any profit
- Liability from decisions is a shared responsibility. The advantage is when the skills and experience of the partners complement each other. The disadvantage is if others are making deals without the knowledge of the others. It is therefore essential to have clear lines of demarcation and communication.
- There are fewer costs and formality in its formation when compared with starting a company.
- It is easier to raise capital because of the numbers of persons involved, but often homes are still used as security.

- Unless covered in the deed of partnership, the partnership can be dissolved if one of the partners dies or goes bankrupt.
- One or more of the partners can cover for holidays and illnesses.

3.3.2 Limited liability partnerships (LLP)

A limited liability partnership is similar to an ordinary partnership, but here the liability is limited to the amount of money invested in the business and to any guarantees given when raising finance. This gives some protection to partners if the business runs into financial difficulty. There is no limit to the number of partners, but a minimum of two must be 'designated members'. The LLP must register at Companies House, and besides the same tax return obligations, as with ordinary partnerships, LLPs are obliged to send their annual accounts to Companies House, which means that unlike ordinary partnerships, their accounts are available for the public and their competition to inspect.

3.4 Limited liability companies (LLC)

Limited liability companies exist in their own right. As the company's finances are separate from the personal finances of the owners, called the shareholders. The shareholders can be individuals, but other companies such as the pension funds and other finance organisations often own the great majority of shares. If the company fails, the shareholders have no liability, but may lose the money they invested as the value of shares drops. The company continues to operate even though the managers and the owners change.

There are two main types of LLCs: private limited company and public limited company. The main characteristics of a private company are that the name of the organisation must include the word limited usually abbreviated to Ltd.; there must be at least one director; it is co-owned by the shareholders; the shares cannot normally be offered to the general public, but can be offered to the employees of the company; and it does not have to declare as much information as public companies.

Public limited companies all have the designation PLC, are usually larger than private limited companies, and must have a minimum of two shareholders and two directors. They can sell their shares to the public provided they have issued shares to a laid-down figure (£50,000 in 2006) before they began trading as a PLC. These are offered to the public on the stock exchange subject to the rules and regulations of the Monopolies Commission.

3.5 Setting up a company

This process is called incorporation and requires two documents. These are the Memorandum of Association and the Articles of Association. The former is concerned with issues such as the company name, where its registered office is located, the purpose of the company and the total value of its shares. A company can only carry out work as stated in its 'objects clause' in the Memorandum of Association. The Articles are concerned with the internal matters of the business such as the rights of the shareholders, the responsibilities of the directors, and the rules governing the annual and any extraordinary general meetings.

These two documents and others which specify the amount of capital invested into the company and the number of directors, are submitted plus a fee to the Register of Companies at Companies House which then issues a 'Certificate of Incorporation'. After this, details are made available for public inspection and private companies can begin trading immediately. Public companies must wait until the Register is satisfied that the necessary share capital has been allotted to the company. A company must have a company secretary.

Profits are normally distributed to shareholders in the form of dividends, but some may be held back for future investments or placed in a contingency fund. How much of these can be diverted from the shareholders will depend on the acceptability to shareholders at the annual general meeting. The Companies Act (1989) is the statute that controls the regulations of UK companies. It includes the rules for producing annual accounts, the directors' and auditors' reports, the appointment of the auditor and the publication of accounts.

3.6 The annual report

The law requires annual reports and accounts. These must contain a profit and loss account and a balance sheet. In simple terms a profit and loss account shows the turnover that has been achieved over the account year less the net operating costs, the difference between them gives the profit before tax. Tax is then deducted to give the profit for the financial period. If dividends are paid, these are deducted leaving the profit that is retained for reinvestment in the business. As shown in Table 3.1, this is usually to show the accounts for the previous years so readers can see if the company has improved or otherwise.

The balance sheet shows the overall state of the business and takes account of assets the business owns and liabilities, such as bank loans and creditors.

Table 3.1 Profit and loss account

	2007	*2006*
	£000s	£000s
Turnover	123.7	108.7
Net operating costs	99.3	88.3
Profit before taxation	24.4	20.4
Taxation	8.3	6.9
Profit for the financial period	16.1	13.5
Dividends	2.0	1.8
Profit retained	14.1	11.7

Table 3.2 Balance sheet

	2007	*2006*
	£000s	£000s
Fixed assets	40.1	38.0
Current assets	50.2	43.2
Current liabilities due within 1 year	25.4	22.8
Net current assets	24.8	20.4
Total assets less current liabilities	64.9	58.9
Liabilities due after more than 1 year	10.3	8.3
Net assets	54.6	50.6
Capital and reserves	54.6	50.6

Table 3.2 shows a typical example. The fixed assets are plant, equipment and buildings. These are adjusted annually to take account of the depreciation occurring over time, although the buildings might become an appreciating asset. The current assets are materials, debtors and any monies held at the bank or in hand.

The current liabilities are those payments to be made within one year and include organisations the company owes money to, known as creditors. The net current assets are calculated by deducting the current liabilities from the current assets and the total assets, less current liabilities, and are the sum of the fixed and current assets less the current liabilities due within one year. The liabilities due after more than one year include long-term finance loans, long-term creditors such as when a supplier is being paid over several years in stages for a substantial order, such as large plant, and making a provision for contingency liabilities for either known future liabilities or the unexpected. The net assets are what is left from the total assets after all the liabilities are deducted.

The capital and reserves comprise the owner's investment, such as the shares either representing the directors' investment in a private company or the shares in a public company sold when setting up the company and any floated during the financial year, and the profit or loss from the profit and loss account. The capital and reserves should equate to the net assets.

There are ratios that should also be looked at to demonstrate the state of the business. These include:

1. *Return on capital employed.* This is defined as: pre-tax profit ÷ capital employed. The pre-tax profit is obtained from the profit and loss account and the capital employed is the total assets less the current liabilities. If the figure is large this means the company has made a healthy profit with little money outlay. If it is smaller than the current interest rates, it would be better to invest the money in a building society or bank.
2. *Profit margin.* This is defined as: trading profit ÷ value of sales, or in the case of construction, work paid for. If this is high then each project completed is making a high profit. If this is low the return on capital employed (1 above) will always be low.
3. *Gearing.* This is defined as: fixed-term loans ÷ net assets. This is a very important way of assessing a business. If it is high, the majority of the profit has to be used to repay loans. If it is low then less is used this way, and if it is zero all loans have been repaid. Future potential lenders would be particularly interested in this figure as it indicates the level of risk and will determine the interest rates at which they would be prepared to lend.

4. *Output per employee.* This is defined as: value of sales or work paid for ÷ number of employees. This is used as a rough guide to benchmark efficiency against the competition. However, since in construction, so much of the work is sub-contracted, it would be difficult to implement in practice.

Larger companies and organisations must also include the following in their annual report:

- Chairman's review – this is an overall summary of the successes and failures of the business over the year, general progress and indicators for the future.
- Auditor's report is a clarification that the company accounts have been produced correctly and honestly, and represent a true picture of the company's financial position.
- Statement of sources of funds – shareholders are interested to know where money has been obtained and that it is legitimate, reliable and the repayment terms are appropriate.
- The way the funds are used indicates to shareholders that the money has been used wisely and in the best interests of shareholders and employees.
- Full record of financial transactions during the period.
- Directors' salaries and bonuses paid are of particular interest to shareholders these days with all the publicity about 'fat cats'.

3.7 Closing a company

If a company gets into financial difficulties and cannot repay its debts it may be closed by a court order requested by one of the creditors who is not prepared to wait any longer for the debts to be repaid and believes a smaller payment is better than waiting longer and receiving even less. This is called liquidation and can be done voluntarily or compulsorily. If voluntarily, then the company has agreed to sell off its assets and pay off the debts. If compulsory, the company has no choice and the court appoints an official receiver. This is usually one of the larger accountancy firms with specialist expertise. They may try to find a buyer for part or all of the business or divide the company into smaller sections and sell them off. Their obligation is to try to get as much as possible for what is left on liquidation.

3.8 Franchising

Purchasing a franchise is a means of taking advantage of an established business and is an increasingly popular trend. The person, or persons, going into the business pays money to the successful business for the franchise. A franchise is a licence to use the name, products and services of the organisation. One of the most well-known examples is the fast-food chain McDonald's. The business also receives the management support systems from the franchiser. The franchise is usually limited to an agreed geographical area and runs for a fixed period, with possible options to negotiate extensions and renewals of contract.

The agreements between the franchiser and franchisee vary depending on the agreement. These agreements will specify the amount and type of management training offered, whether the franchisee has to find the funds to start the business or if the franchiser will loan the money. Profits made may have to be shared and it is normal for the annual accounts to be shown to the franchiser. The extent of each other's liability is also covered in the agreement. The main features are:

- There is less outlay and risk due to the support normally given. The risk increases as this level of support decreases.
- The product or service has already been marketed, proven and accepted by customers.
- Franchisers may provide national or regional advertising as part of the franchising package.
- There can be a conflict between what the franchiser wants and the best interests of the franchisee.

3.9 Co-operatives

These comprise a voluntary association of people called members. The most famous one being the co-operative movement started in Rochdale in 1844, set up by 28 weavers, referred to today as the Rochdale Pioneers. Their idea was to purchase foodstuffs at wholesale prices and sell them to the members at the market price. The profits were then divided to members in proportion to the amount they had purchased in the year. As late as the 1970s regular shoppers at the Co-op shops could still become a members, collect their dividend based on the amount spent each year and vote at the annual general meeting. However, other than when using the Co-operative Bank for clients who are formed as a co-operatives, it is unlikely that a co-operative business would be set up to construct or design buildings. Since readers may be involved at some stage of their lives in voluntary co-operative organisations such as non-profit making clubs and societies, they are discussed here.

The main advantage of a co-operative is that members support each other. The aim is not necessarily to make a profit, but not a loss either. Each member is entitled to a vote irrespective of the amount of investment. Today the Co-operative Bank continually consults their customers about business ethics and has adapted their philosophy to reflect members' interests and concerns which is shown in their attitude to such issues as environmental concerns and arms trading. In a co-operative, the members democratically elect the manager of the co-operative. Limited liability can be obtained by registering with the Register of Friendly Societies. The main features are:

- Anybody that works in the organisation can become a member or is automatically one.
- The business of the organisation is carried out for the mutual benefit of all the members. All profits are shared if the organisation is a profit-making one.
- Each member has a vote and the members elect management.
- The organisation must be socially aware and act responsibly to others including, customers, suppliers and the local community.

3.10 Board of directors

Although shareholders own limited companies, it is not their responsibility to run them. They employ directors to do this on their behalf and, if not satisfied with their performance, can vote them off the board at the annual general or extraordinary general meeting. The board is responsible for setting the strategic objectives and making sure there are sufficient resources in place to meet them. They are responsible for all matters concerning the running of the company including health and safety, tax and employment law. They have a duty to ensure the financial accounts are accurate and honest. Indeed, it is a criminal offence to do otherwise. The board members take collective responsibility for the overall performance of the company.

There can be executive directors and non-executive directors. The former are full-time employees of the company and will have specific responsibilities for certain aspects of the business, such as director responsible for finance, human resource management, contracts and so on, as well as their board responsibility. The non-executive directors are appointed because of their specialist knowledge and expertise or sometimes their contacts or image and standing within the community. If the business operates overseas then a retired foreign secretary could access influential politicians in another country. A member of the extended royal family gives a feeling of stability and integrity to the business. Their role is also to challenge the ideas put up by the executive directors as they are outside the normal day-to-day thinking within the business and are able to approach matters with a fresh and unbiased point of view.

The board must work within the limits set down in the Memorandum and Articles of Association which may have laid down certain restrictions on how and where the business operates, the type of work to be carried out and whether or not money can be borrowed. They must treat all the shareholders equally irrespective of the number of shares owned. If they have any conflict of interest this must be declared. This can often occur with non-executive directors who may have interests in other companies. They must obey the law and occasionally can be responsible for the actions of their employees, such in matters of safety where, if they have not set up standard procedures and effective monitoring, they could be accused of negligence. Finally, they must not make personal profits at the company's expense, like selling or buying company shares when they have privileged information.

The board will meet at regular prescribed intervals. The frequency will depend on the business itself. It would normally be anything from once a month to once a quarter.

3.11 The chairman of the board

The chairman of the board has different responsibilities from the chief executive (section 3.12). The chairman or chairwoman, sometimes referred to as the Chair, is, as the name implies, responsible for chairing the board meetings as distinct from running the company. Their role normally includes:

- setting the agenda after consultation with the chief executive;
- ensuring the board runs effectively and all items on the agenda are given appropriate time for discussion;
- concluding each agenda item with a course(s) of action when necessary;
- ensuring through the company secretary that all the papers for the meeting are distributed to give sufficient time for proper consideration, and the subsequent minutes are promptly sent to members;
- ensuring there is proper communication from the executive to the non-executive members about the running of the company;
- ensuring there are appropriate checks and monitoring procedures in place so the board can properly evaluate the success or failure of the business.

3.12 The chief executive

The role of the chief executive, or managing director, is to run the business on a day-to-day basis on behalf of the shareholders. Their responsibility includes:

- developing and delivering the strategic objectives, which have been agreed by the board;
- giving responsibilities to senior management to carry out this work to meet the objectives and monitor their performance;
- preparing the annual budget and financial plan for the business and establishing medium-term financial projections;
- recruitment, developing and retaining good quality staff;
- establishing and monitoring risk-management strategies;
- reporting to the board with accurate and timely information so the board can discharge its responsibilities properly;
- consulting with the chairman of the board on all significant matters;
- representing the company on matters pertaining to the business of the company.

References

Business Link (2008) Home page. Available: www.businesslink.gov.uk. Accessed 14 October 2008

Freeman-Bell, G. and Balkwell, J. (1993) *Management in Engineering*. Prentice-Hall.

CHAPTER

Strategy Planning

4.1 Introduction

Companies evolve over a period of time; expanding, diversifying and contracting, depending on circumstances. There is a need to periodically re-evaluate the business, not just because there is a problem, but because the business world is always changing and the company needs to position itself to take full advantage of opportunities that might arise. Sometimes the changes are small, but in other cases the changes are dramatic moving the business in a different direction with major implications to the employees and their future with the company. Examples of this are when in the 1990s Wimpey and Tarmac decided to exchange parts of their business, Wimpey giving its contracting arm to Tarmac and in return received Tarmac's house building. In 2001 Alfred McAlpine sold its house building and used the proceeds to build up facilities management and utility services.

Strategic planning has become increasingly important over the last few decades because both the external and internal environments have become much more complex, and the techniques available to management have become much more sophisticated. It takes place in both the private and public sector. In essence the company is saying 'where do we want to be in five years time and how do we get there?' This requires senior management decisions, the ability to allocate large sums of money, if necessary, and to understand the company's interaction with the external environment and the implications with the internal environment. It is done to identify and achieve competitive advantage, to give the business direction all the employees can identify with, which can encourage innovative ideas to achieve the goals set. Once the strategy has been decided it will fall into one or more of the following: expansion, contraction, diversification or consolidation.

4.2 Levels of strategy

Depending on the size of the company, there are three levels of strategic development, as shown in Figure 4.1. Single businesses will only have two.

The group corporate-level strategy determines which of the company's businesses will continue to operate, how resources will be allocated within the group, and whether any of the third level will be co-ordinated for the group as a whole. For example, research and development may be carried out on behalf of all the businesses. It will also determine the amount of growth and profit expected from each of the businesses and their contribution to the group finances.

Business-level strategy looks at the requirements for each of the businesses assuming each is autonomous and has its own competitors. In the case of a national and international construction company, this could be a UK operation, a European operation, house building, general contracting or civil engineering. Within the UK, it could be a northern and southern operation or smaller areas such as the north-west, Scotland or the south-east. The strategy will consider the amount and level of competition, its objectives and the opportunities within which the business is operating.

Functional-level strategy looks at how each functional area in the business needs to perform to meet the overall strategy for the business. For example, if the business is expected to expand by a given amount the marketing strategy has to be geared to make that happen. The human resources section has to

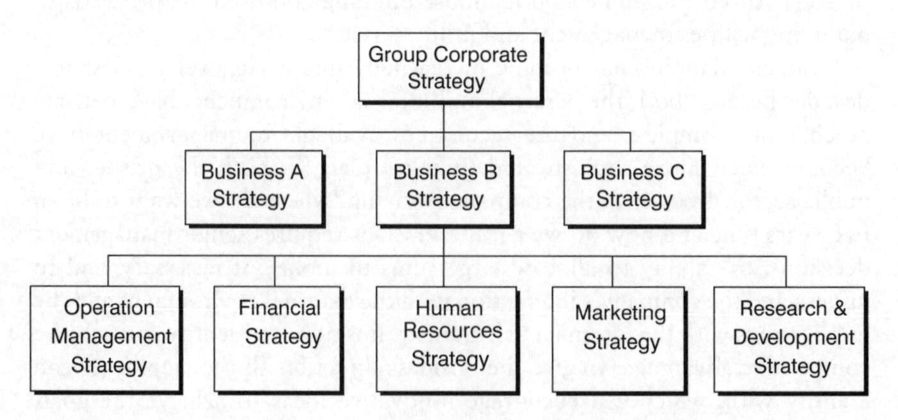

Figure 4.1 Levels of strategy

take steps to recruit and train the required number and type of personnel. If the business is to shrink, they must develop a redundancy strategy to have the number of staff to meet the new demands, or if the business is to diversify recruit personnel expert in the new work of the business. The group needs to co-ordinate the strategies across all the businesses to maximise efficiency and to ensure the overall corporate objective is being achieved.

4.3 The strategic management process

The components of the strategic management process are demonstrated in Figure 4.2. The first stage is to identify the company's current mission, strategic goals and objectives. The strategy may be modified as a result of the exercise.

The first stage is to identify the company's mission: what business is the organisation in, where it is heading and when does it wish to achieve its objectives? Using the example of higher education, is the university in the business of training students for particular jobs or is it in the business of producing graduates with a well-rounded education so they are adaptable to the work situation? Does the university wish to take the less able students and add value using high levels of staff involvement or take the higher ability students and expect them to read for the degree on their own with few lectures and individual tutorials, as at Oxbridge?

In construction, questions could be: is the business to construct speculative housing and/or general and civil engineering contracting or a particular part

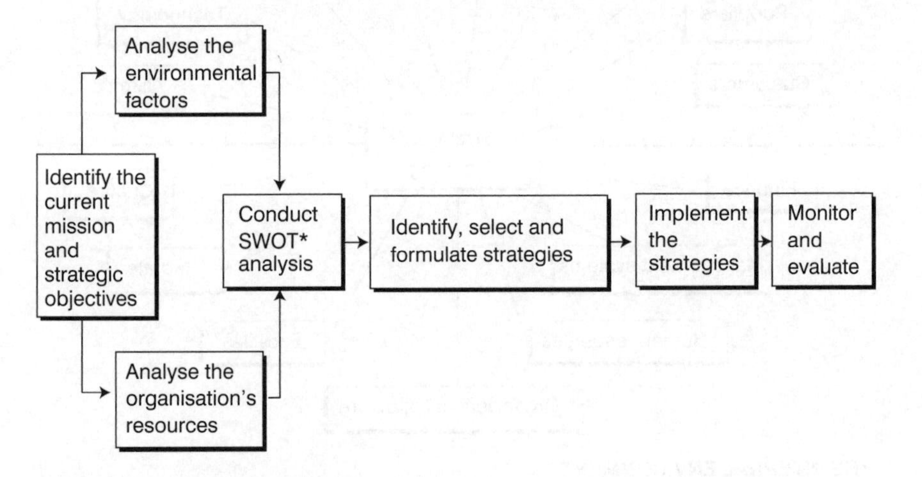

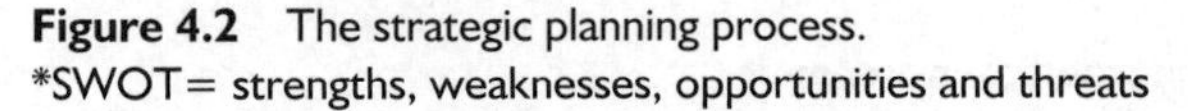

Figure 4.2 The strategic planning process.
*SWOT= strengths, weaknesses, opportunities and threats

of the market? Is the business one that reacts to tender enquiries or is it one that is active in generating work? Is the company concerned with providing a quality service or product? Does the company want a zero accident culture? Does the company wish to move to being known as 'an environmentally friendly' business? Does the company wish to minimise the time spent making good faults after handover?

Many companies lose their way when their mission objectives are unclear or wrong. There are well-known examples in the retail trade where customers drift away, because the shops no longer provide them with what they want and they go elsewhere to those that do. The railway business in the USA had a problem because they saw their business as being the railway business rather than transporting people and products.

The second stage is to assess the environment in which the business is to operate and the resources within the organisations, as summarised in Figure 4.3. Without this analysis it would be difficult to plan the future of the

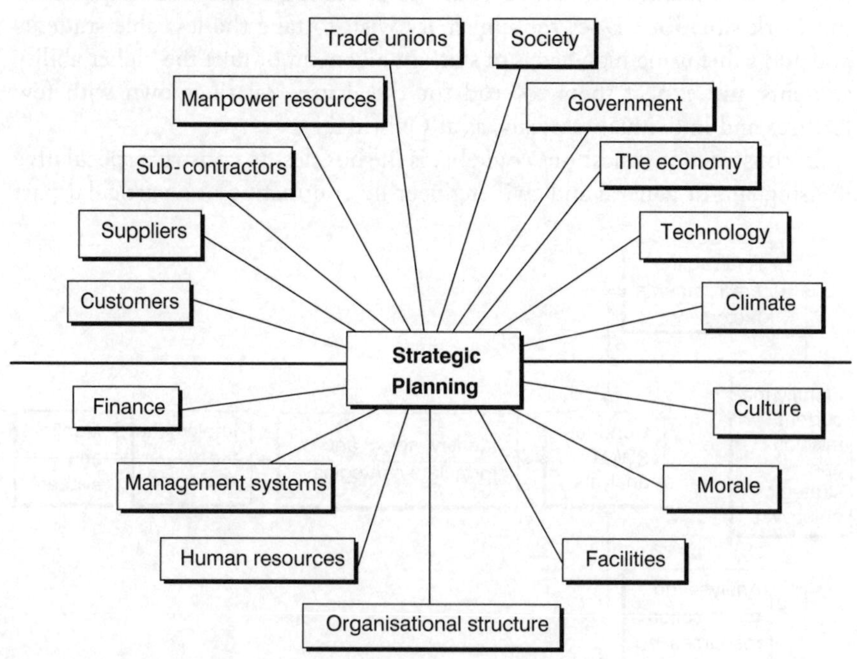

Figure 4.3 The internal and external environment

company. Customers, present and potential, are the crucial element of any analysis of the external factors affecting the development of a strategy. The status of suppliers of materials and components determines the opportunities for expansion. If they are over-stretched, new ones have to be sought out and, if not available, this will affect any decisions made. Similarly, with sub-contractors, whether they can meet the demand and quality required. The availability of adequate trained manpower, both manual and cerebral can impact significantly on the business, especially if one of the requirements for working in a particular area is that local labour has to be employed. Trade unions have become very pragmatic and work with employers for the good of the industry and their members. The days when, for example, many companies would not work in Liverpool because of the militant trades unions have now gone. There are parts of the world where the way labour is organised and controlled needs to be understood, as it can affect the prospects as, for example, in Hong Kong where much of the labour is controlled by the Triads.

Social changes have an impact on the construction industry, for example, in housing: the increase of one-parent families and the need for social housing. Society can also determine other changes in requirements, as did the revulsion towards multi-storey housing as a solution to housing shortages in the 1970s; a trend now reversed with many multi-storey flats being constructed in inner cities. Environmental pressures on where construction occurs and the types of buildings are other examples of societal influence. Government (local, national and European) determines policies and enacts laws that influence the marketplace. Nationally, the Private Finance Initiatives are a good example of this and the current push for affordable housing another, as this may result in the relaxing of planning regulations and the availability of previously protected land. The European Community (EC) produces directives affecting employment and free trade, hence public contracts over 5 million euros have to be advertised across the EC. Contractors need to be aware of the economic climate and future trends. This is affected by national governments and world events since we now exist in a global economy. Technologies are changing fast, not just impacting the way construction takes place, but also in the requirements of the clients both in the design of buildings to accommodate the new technologies and to cater for the way their business operations change. Finally, climate can have a significant impact on the construction process depending on the part of the globe in which the company decides to operate.

Internally, the financial state and ability to borrow money to finance further operations needs to be analysed. Are the management systems in place appropriate to the organisation's needs or should they be modified or improved? The human resources department will assess the type of manpower available within the company and the match with the strategic plan to establish whether or not training and promotion from within will cover the needs, or whether recruitment from without is necessary. Organisation structures can become ineffectual and changes must be made. Offices may be inappropriate with poor internal communications as a result. The buildings housing staff may be outdated, the wrong size or in the wrong geographical location. More difficult to establish is the overall morale of the staff. If it is low then radical changes to the organisation may be difficult. The culture needs also to be explored; changing this takes time and needs to be managed carefully. The culture may have been determined by many things, such as the business previously carrying out traditional contracting and is now necessary to move into the management contracting arena.

4.4 SWOT analysis

SWOT is the acronym for strengths, weaknesses, opportunities and threats. The strengths and weaknesses are concerned with the internal environment of the company, whereas the opportunities and threats are factors affecting the company from outside. Typical issues that might be explored in a SWOT analysis are shown in Table 4.1.

It is normal practice to quantify the relative merits of each of the four categories and prioritise them. By carrying out a SWOT analysis, management has a much clearer understanding of where the company is and the environment in which they operate. This gives the opportunity to address the weaknesses and use the strengths to overcome the threats and explore and take advantage of any opportunities.

Whilst Table 4.1 gives indicative issues that might be raised, in practice the issues evolve, through senior management brainstorms or by consulting key staff and asking them to list their views on the four categories. These are then collated and a list produced for each of the categories, prioritising the responses from those consulted.

One of the major issues will always be competition from other companies and looking at this issue can enhance the SWOT analysis. Michael Porter, a strategy expert, developed his 'Five Competitive Forces Model', comprising rivalry, bargaining power of customers, bargaining power of suppliers, threat of new entrants and threat of substitute products or services. Rivalry is the way competitors use varying tactics to obtain work, by offering discounts, reduced construction periods, advertising and various incentives in the housing market such as free legal fees, stamp duty paid and carpets

Table 4.1 Typical issues for consideration in SWOT analysis

Strengths and Weaknesses	*Opportunities and Threats*
Financial status	New markets
Competitive edge	Exiting customers
Image	Competition
Strategic direction	National economic revival or decline
Facilities	Labour availability/redundancies
Management expertise	Suppliers workload
Market leader	Sub-contractors workload
Cost advantages	Expansion of the European Community
Reputation for quality	Inflation levels
Reputation for completion on time	Terrorism and war
Construction expertise	Mergers and takeovers
Staff turnover	Location of the business
Land bank	Employment legislation
Morale of staff	Risk

throughout. This gives an indication of how much competitors are spending which can affect their profit margins.

The bargaining power of customers is how strong they are in being able to force prices down or obtain an improved service for the same price. This is a function of the market forces of the day or the strength of their order book. The stronger the bargaining power, the lower the profit, except in the case of serial or term contracts (*Finance and Control for Construction*, Chapter 8), where there are other advantages because of the continuity and familiarity with the work.

The threat of new entrants in construction can come from overseas and from existing contractors who have decided to diversify into a particular type of work. For example, when the workload of the industry declined in the 1980s, some of the larger contractors diversified into smaller works. This inevitably puts pressure on profit margins as companies strive to maintain their market share against the new competition or reducing market.

The threat of substitute products or services can be interpreted in construction where new forms of contracts are used that require a differing approach in attitudes and culture. The changes can be significant as shown by the increase in design and build, management, and construction management contracts and the Private Finance Initiatives. The threat of substitute products is of more concern to the industry suppliers such as with uPVC replacing timber for window frames.

As a result of all these analyses and the SWOT analysis, the mission and objectives of the organisation may have to be amended.

4.5 Identify, select and formulate the strategies

The strategies selected need to be set for the three levels of the business: corporate, business and functional, as indicated in Figure 4.4. There are two accepted approaches to this: grand strategies framework and the corporate matrix strategy.

4.5.1 The grand strategies framework

Sometimes referred to as the master strategy, as demonstrated in Figure 4.4, it comprises three main approaches: growth, consolidation and retrenchment, sometimes referred to as defensive.

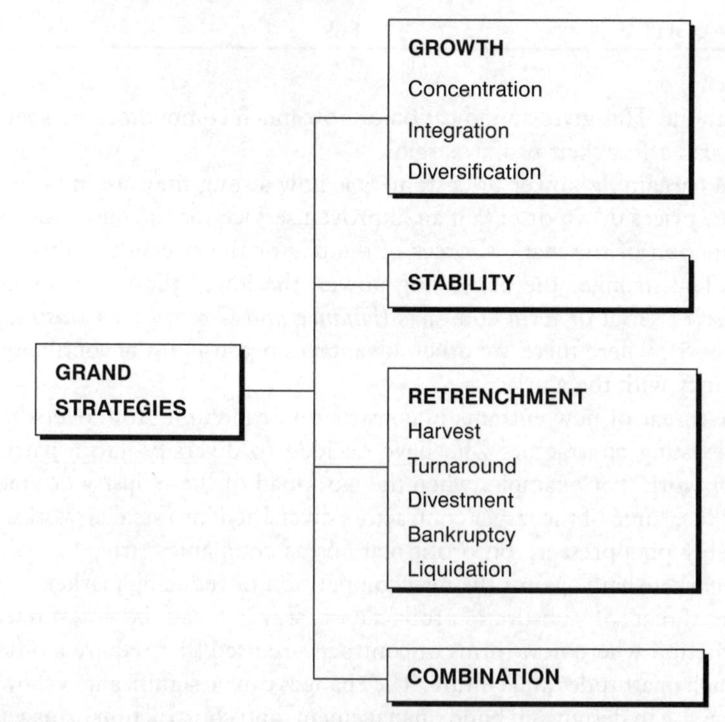

Figure 4.4 Grand strategies (adapted from Bartol and Martin (1994) © The McGraw-Hill Companies, Inc.)

Growth strategy: A growth strategy, as the name implies, means employing strategies that expand the business in some way. The first, concentration, is either to direct effort into a part of the business with the aim of securing a larger market share, or to develop a new product in the existing business, for example, in speculative housing by using timber frame as well as or instead of masonry construction.

Integration is taking on work in the supply chain currently carried out by others, or not producing for a client, but selling directly. This is interesting in relation to construction as the UK industry has done exactly the opposite in the former case, and in the latter, other than in the speculative market, has always built 'a product' for a client. The reason for sub-contracting work has been one of cost savings and the ability to call in labour as required rather than having to find work and guarantee continuity of employment in the temporary and changing nature of construction projects. There seems little likelihood of this reverting to the scenario of employing labour directly. However, it would be interesting to explore the possibility of becoming the developer as well as the contractor as a means of growing the business.

The third approach is diversification into new markets. The advantage of this is the business is less exposed if one of the other strands comes under competitive or economic attack. However, there are risks involved when entering any new market as it unfamiliar territory. In construction, this could mean moving into the PFI marketplace, offering high-quality, exclusive executive speculative homes as distinct from the general market, carrying out the refurbishment and maintenance contracts or starting a facilities management operation to complement the PFI work.

Stability strategies: This is a situation when the company decides it does not want any further growth and will concentrate on consolidating its business. This can occur when a business has grown too fast and feels it is over-stretched resulting in unsatisfactory service to its customers so a period of stability is required. In a declining marketplace maintaining its place in terms of market share can also be considered a stability strategy.

Retrenchment strategies: Sometimes called a defensive strategy, as some companies dislike the term retrenchment, it is designed to reduce the company's operations normally by cost cutting measures or by reducing the company's assets.

A harvest strategy is used when the company intends to move out of a market, but in the meantime has to carry on until the existing order book has been completed. This requires as minimal investment in the business as possible, while trying to maximise profits and cash flow. Examples of minimising investment would include not decorating the premises, not investing in new plant and equipment, and ceasing manpower training unless retraining for a new occupation elsewhere in the group.

A turnaround strategy is employed when the company is in decline and the aim is to reverse this trend and restore the once-achieved profitability levels. This will involve attention to all the cost centres and usually involves cost-cutting measures, including redundancies or selling off the non-profitable parts of the business. The latter is known as divestment.

Bankruptcy is applicable to individuals who can no longer pay off their debts and cannot be pursued by their creditors once the courts have allocated any remaining assets they might have. Companies go into either compulsory or voluntary liquidation. The former is when a creditor goes to court and the court appoints an official receiver to sell its assets and pay off its debts, the latter is when the company agrees to sell its assets to pay off its debts.

Combination strategies: This is when different parts of the business follow more than one of the above strategies where, for example, one part of the group is pursuing a growth strategy and another is contracting.

4.5.2 Corporate portfolio matrix

The grand strategies approach looks at the company's overall direction, whereas portfolio strategy aims to help managers determine the types of business in which the company should be involved. The Boston Consultancy Group (BCG), a well-established consultancy organisation, developed it in the 1970s. It comprises four cells, referred to as stars, question marks, cash cows and dogs in the top left, top right, bottom left and bottom right respectively. The horizontal axis expresses the percentage of the market share and the vertical axis the anticipated growth relative to growth in the economy as a whole. In the matrix squares are used to represent the size of the individual business as a percent of revenue relative to the other businesses in the company as a whole.

The businesses in the star category are high growth and have an anticipated high market share. This is a rapidly expanding market and the business may require a significant investment that is greater than they can earn. Those in the cash cow quadrant have a high market share in a low growth market. They usually generate high returns but have only a limited opportunity for future growth. Those with the question mark have a low market share, but in an anticipated high growth market; in essence there is a question mark over them as they are high-risk ventures that might or might not be successful. They will require substantial investment if it is decided to support these businesses. Dogs are the low growth and anticipated low market growth area. They produce little profit, but do not require much investment. These need either to revitalised as in harvest and turnaround, or divested or liquidated (section 4.5.1).

Construction businesses do not have as many separate companies as some of the large manufacturing groups, but this method can be used to look at the relative performances of the regions of a national and international construction company and may indicate, for example, two or more regions should be merged into a larger region. It should be remembered this is a guide to assist senior management to make strategic decisions and should not be taken in isolation. Each separate business should be analysed independently as there may be good reason why the business is shown in a particular quadrant at that point in time.

4.6 Formulating business-level strategy

This is concerned with the way a business within the group competes and operates (section 4.1). In the construction industry this usually will be at regional and organisational levels. This section will explore two approaches: adaptive strategies, developed by Raymond Miles and Charles Snow; and competitive strategies, developed by Michael Porter.

The adaptive strategy identifies four strategy types: defenders, prospectors, analysers and reactors. Defenders produce only a few products or services aimed at a narrow part of the market, and work very hard to try to keep others out of their 'patch'. If successful they maintain a niche market other competitors find it difficult to penetrate. Prospectors look for innovation and believe their success will be founded on finding new ideas and exploiting that market. To achieve this they need to have organisations that are quick reacting and prepared to accept and absorb change. Analysers aim to minimise risk and maximise the opportunity for profit. They copy good innovative ideas that prospectors have developed. They are more efficient than prospectors, but because they take less risk, produce smaller profits. The reactors behave inconsistently and tend to perform badly. They are reluctant to follow any strategy aggressively and usually lack focus.

Competitive strategies consist of three alternatives: cost-leadership strategy, differentiation strategy and focus strategy. The common requirements for each of the three strategies are:

- A cost-leadership strategy is when the company sets out to be 'the' cost leader in the business and not just one of the contenders. The product or service provided must be seen as being similar to the rivals. To be a cost leader, and have cost advantage, the company needs efficient operations, innovation, low labour costs (this does not necessarily mean low wages), economies of scale and access to suppliers who give preferential rates. All has to be achieved without affecting standards of quality, safety and environmental issues. To do this requires very tight controls.

- A differentiation strategy is about producing products and services unique to the industry. These differentiation factors could be high-quality, innovative designs, quality of service, speed and reliability of completion, very low accident rates or a combination of several of them. 'What makes us different?' has to be marketed as a concept. Both the cost-leadership and differentiation strategies are aimed at a wide range of the market.
- A focus strategy is aimed at a very narrow band of the market by specialising, and is probably best suited for small businesses. The segment of the market could be a geographical region, a type of customer or a specific part of the market, for example a structural steel frame erector. The strategy still requires either a cost-leadership or differentiation approach.

4.7 Strategy implementation and monitoring

Once the strategy has been formulated it has to be implemented. It is essential that any implementation should be carried out throughout the business in all departments and functions at the same time. Galbraith and Kazanjian highlighted five principal factors that should be considered: technology, human resources, reward systems, decision processes and structure.

Technology is defined, in this case, as the tools, equipment, work techniques and knowledge used by the company to deliver its product or service. There has to be a balance between the technologies available in matching the strategic objectives. So, for example, if the strategy is to reduce costs, new technologies may have to be employed to achieve the strategy.

The organisation needs the right human resources with the appropriate skills employed in the correct positions, both geographical and functional, and unless this is in place, the strategy cannot be accomplished. Once it is, there are further advantages which affect the company's competitive edge because the employees are more likely to be able to be more productive and innovative in their approach to work.

Reward systems are not just about bonuses, but also concern promotion prospects and job satisfaction, resulting in higher motivation and less staff turnover. However, linking bonuses to achieving the company's strategic objectives is important especially if designed in a creative way. For example, if one of the objectives is to reduce the number of accidents in a region of a construction company, linking the regional directors' annual bonuses to the safety record of their patch would certainly focus their attention and that of the subordinates.

Decision processes is the ability to resolve questions and problems that occur within the business and is, in part, linked to the communications systems in place and the culture of the organisation. A frightened subordinate is less likely to have an open discussion with a supervisor than one who isn't, and a dictatorial supervisor will have little interest in the opinions of those they manage. There is also the issue of having good decision-making processes in place when developing the strategic plan, otherwise how do you know that you have the best plan and have covered for all eventualities?

The organisation structure is the formal way in which the various parts of the company interact, be it horizontally or vertically, sometimes referred to as the organisation chart. It needs to be designed so the different parts of the business can co-ordinate their work to meet the strategic objectives.

Finally, once in place management needs to periodically monitor the key factors that affect the achievement of the strategic objectives to ascertain whether or not the implementation is working and then to make adjustments depending upon what is found. Sometimes the strategic change is seismic in proportions and management has to change vary carefully (Chapter 7).

References

Bartol, K.B. and Martin, D.C. (1994) *Management*, 2nd edn. McGraw-Hill.

Carnall, C.A. (1997) *Strategic Change*. Butterworth-Heinemann

Galbraith, J.R. and Kazanjian, R.K. (1986) *Strategic Implementation: Structure, Systems and Process*, 2nd edn. West Publishing.

Hussey, D. (1994) *Strategic Management*, 3rd edn. Pergamon.

Megginson, L.C., Mosley, D.C. and Peitri, P.H. (1989) *Management Concepts and Applications*, 3rd edn. Harper Row.

Miles, R.E. and Snow, C.C. (1978) *Organizational Strategy, Structure and Process*. McGraw-Hill.

Porter, M.E. (1980) *Competitive Strategy*. Free Press.

Robins, S.P. (1994) *Management*, 4th edn. Prentice-Hall.

CHAPTER

5 Marketing

5.1 Introduction

The construction industry is different from most other industries because it provides a bespoke service producing a product which the client has commissioned that, generally speaking, is different in every case, the main exception being the speculative housing market. The concept of 'selling' is therefore alien to many in the industry except to those in the speculative market. At first glance, the industry reacts to clients' wishes by tendering for work and if successful, constructs the building or provides the infrastructure. However, whilst traditionally this was the case, the construction company had then to contend with getting their name on preferred lists so they could tender, and more recently make presentations to be selected for projects and serial contracts. This has made the industry more conscious of the need to market and sell their services.

The Chartered Institute of Marketing defines marketing 'as the management process responsible for identifying, anticipating and satisfying customer requirements profitably'. Examples of these customers needs in the speculative marketplace, include: what size and location houses should be, the quality of finishes, and so on. In mainstream contracting, as contractors have become more involved in the design processes, they are in discussions with clients to establish what the clients really need; they apply valuation engineering techniques; there is greater emphasis on on-time completion and less on using the contract clauses as reasons for not doing this to claim extra payments; and use total quality management to ensure customer satisfaction.

There are various stages of marketing for the construction industry which can be briefly summarised as follows:

- market intelligence systems and market research
- the customer
- the external environment
- forecasting
- market positioning
- new products or services
- distribution decisions
- advertising
- other forms of promotion
- market planning.

5.2 Market intelligence systems

The success or failure of the marketing process is dependent upon the quality of the data obtained from market research and information it provides. The quality of the management decision is only as good as the facts available at the time. Day-to-day management does not have the time to spendcollecting and examining large quantities of information, but rather has to make rapid decisions based on clearly important information available at the time. However, making decisions about the long term is a different matter and marketing falls into this category. The role of market research is to minimise the uncertainty as far as possible by providing certain information, known as market intelligence. Data are the facts, information comes from data that have been selected and sorted for a specific purpose, and intelligence is the interpretation of the information after analysis.

Information technology allows marketers to to collect, store and organise data. These systems are called marketing intelligence systems (MIS). The information is obtained from three prime sources as shown in Figure 5.1.

In construction, internal data related to performance is primarily obtained from the estimating, purchasing and contract management

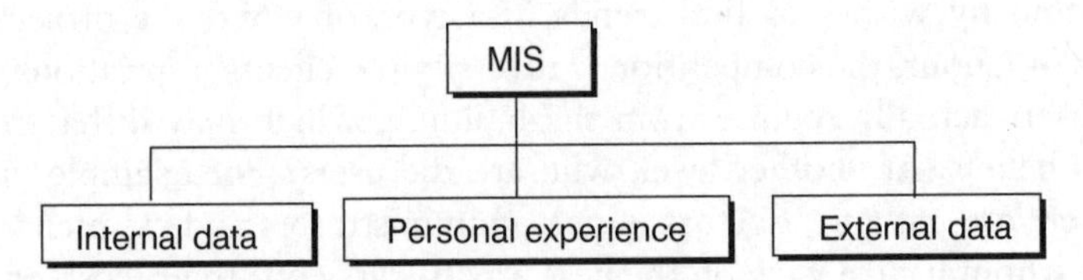

Figure 5.1 Sources of data for market intelligence systems

departments which give information on costs, both actual and estimated, success rates in estimating, competitors' successes, completion times, durations of activities, names of the clients and design team members. Further internal data can be derived from written reports, aide memoirs and by questioning staff.

One should never underestimate the knowledge gained from personal experience. For example, competent site managers and estimators can look at a set of drawings and in relatively short time advise on the approximate cost of the project, its duration and when key stages in its construction are likely to be achieved. Senior managers have an instinct and feel for the marketplace resulting from their contacts with clients, existing and potential. Purchasing managers are good at predicting by how much materials suppliers and subcontractors are likely to reduce their quotations if the contract is awarded.

External data can be collected from the web, providing the person doing so is able to discern the useful material from the rubbish. However, data about clients and their businesses are generally sound. It is often possible to discover a lot about other organisations from their web sites. Companies have to lodge their annual accounts with Companies House, another source of data. The news media is an excellent source of information. There are the broadsheets, but also the specialist newspapers such as *The Economist*, and the trade press. Libraries, professional institutions and trade associations are excellent sources, especially for directories, with their staff often eager to carry out searches. There are an increasing number of computer databases that can be accessed, usually for a fee or subscription.

5.3 Market research

Market research is the process of collecting information from consumers, the users of the building in the case of construction, and the clients. Sampling is normally used because questioning all clients and users would be too large a task. It can be carried out by the company's staff if qualified, or by employing specialist market research firms.

The first stage of the process is to define the objectives of the research: why is the research being done and what to we want to gain? It could be that the company wishes to find trends in a type of work, the projected availability of labour, the competition if diversifying, clients' aspirations, or what the users actually require from the building, which may differ from the client's beliefs. At another level, who are the users? For example, in a hospital there are patients, doctors, nurses, administrators and maintenance staff, all of whom have a view on the best way to carry out their work or be treated.

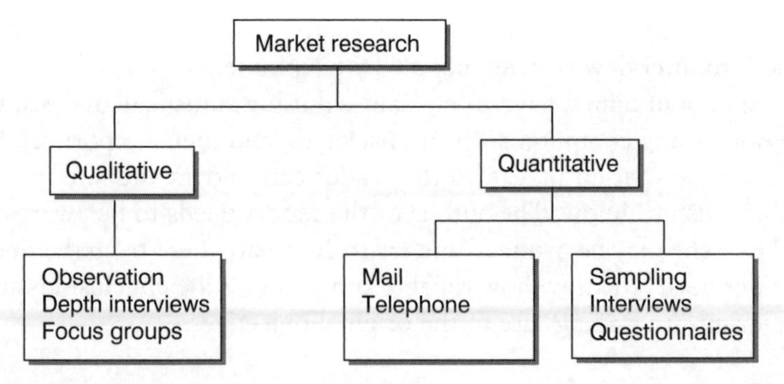

Figure 5.2 The research method selection

The second stage is to plan the research. How much depth is the study going to go into, as the deeper the questioning, the more expensive it will be. It can vary from identifying problems and determining possible courses of action, to conducting detailed surveys.

The third stage is to select the method of research (Figure 5.2). These can be defined as the qualitative and quantitative approach. Qualitative research discovers people's attitudes, opinions, ideas and work methods. This is achieved by several methods. One is by observation of the people in a building and seeing how they react. For example, watching patients in the outpatients department in a hospital and noting how they react, how much they move about, and their ability to find toilets and the doctors. This is linked to method study in the work-study section (*Operations Management for Construction,* Chapter 3). In-depth interviews are conducted usually by means of a structured interview, where all the questions are thought out before the interview and answers solicited in a planned sequence, or by using semi-structured interviews where the main questions are planned, but are more open-ended to allow more freedom by the respondent. These are more complex to analyse afterwards. Another method is focus groups, comprising 6 to 10 people who are encouraged to discuss and debate the answers to prepared questions. The interviewer is there not just to pose the questions but also to act as the chairperson to encourage everyone to contribute.

The quantitative approach aims to obtain enough data for a statistical analysis. These are conducted using questionnaires. There is a science in producing questionnaires which should be studied before composing one, and the reader is advised to look elsewhere for this such as, Hague's *Questionnaire Design*. It is important to choose the sample carefully to ensure it is representative. It is also important to obtain as high a return response as possible otherwise the statistical validity can be questioned. These are difficult tasks. Once the questionnaire has been formulated it can be sent out to the respondents for completion. Alternatively, it can be used as it is, or modified, to interview selected people by telephone.

All the data obtained have to be analysed using statistical analysis, which can be done using computer software packages, and then a report produced. This should be written in a form the reader can understand and give them what they wish to know. The author of the report needs to be aware of any biases he or she may be prone to and try to eliminate these from the findings. The reader needs to know how reliable and accurate the conclusions are and how relevant the findings are to the original objectives.

5.4 The customer

The construction industry serves a wide range of customers including national government, local government, hospital trusts, commerce, industry, the retail sector, housing associations, individuals who either commission work or buy from the speculative sector, and users of the building who may not necessarily be the owners, such as visitors and employees. It provides new buildings, maintains and alters existing buildings, provides a complete facilities management service and then demolishes the building after it has served its useful purpose.

Deciding whether or not to have a project constructed or modified is different from purchasing a manufactured product in many ways, depending on who the customer is. For simplicity the customer is considered in three broad categories, the public sector, the private sector and the individual. The prime determinates for the public sector are the government and local councils which determine the type and use of the project, both building and civil engineering, and the amount available for spending. In recent years the amount spent has also been augmented by the use of the Private Finance Initiatives where private sector money has been sourced. The prime marketing position to obtain this work is on the approved list for tenders and the ability to put together bids and presentations. It requires market research to anticipate trends and changes in political leanings. Good networking is also useful.

The private sector and individual customers are more complex as it is necessary to understand the nature and requirements of the customer. Figure 5.3 demonstrates some of the factors that can influence customers. When considering these factors, it is important to be aware that they continually change as society and the world changes.

Economic influences are key issues in any decision to build, usually determined by government policy and the current and perceived future of the country's economy. How much government is prepared to spend and

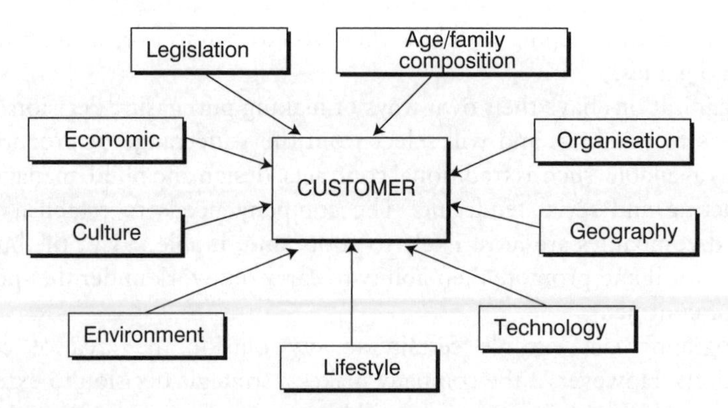

Figure 5.3 Factors that influence customers (adapted from Czinkota *et al.* (1997) with permission of Wiley-Blackwell)

how much it passes to local authorities is determined by the Chancellor. The public sector takes a longer economic view and considers likely trends and the influence of global economics. The individual who purchases or extends a house is concerned with residual income, interest rate trends, any government policy affecting house purchase such as levels of stamp duty, homebuyer information packs, and the ability to raise finance from the lenders, as well as the cost of property. Older couples are using downsizing as a means of releasing capital for their retirement. The decline of company penions will in the long term have an impact on the housing market.

Legislation from time to time creates construction opportunities, although not always from a continual need to reform. Resulting from the tragedy at Hillsborough in 1989, and the subsequent Taylor report, there was a major change and requirement for the design and construction of sports stadiums. Part M of the Building Regulations, 'access and facilities for disabled people' is another example of how construction work has been generated.

The age of customers in house purchase is a serious issue as first-time buyers are becoming older as property prices increase. The size and composition of families is changing with an increase in single-parent families, fewer children per household and with people living longer resulting in more elderly couples or single habitation. Younger purchasers have higher expectations than their parents who were prepared to wait until they could afford to purchase things they aspired to. Younger purchasers are more likely to expect the house to have everything they need when they move in. The choice and style of the design of fitments, such as kitchen units, can be related to the age of the purchaser. The manner in which a show house is presented could be different depending on to whom the houses are aimed. For example, if for families with younger children, then the children's bedrooms would be filled with furniture in keeping with how they would wish to live, with up-to-date toys and games.

Organisations have their own ways of making purchasing decisions when it comes to buildings and will select from the wide range of procurement options available, such as traditional contracts, design and build, management contracting and serial tendering. The company needs to establish which method companies are most likely to adopt and, if able, adapt the business to cater for it and promote their ability to carry out work under the specified form of contract.

Geography can exclude certain activity outside the area of current operations. However, if the company makes a strategic decision to extend its area of working, market research could be necessary to break into the new area. Some customers operate nationally and may wish only to have to deal with one office even though the work is in another area of the construction company. For example, much of Marks & Spencer's work was carried out by Bovis Lend Lease's southern office and the northern director watched from his office in Manchester as the new M & S prestige store was reconstructed after the IRA bomb by the southern division. In the housing market, identifying areas people would prefer to live in can be determined by the school in the area, facilities provided and transport access, not to mention attractive areas for holiday homes, although to date this is provided mainly from the existing housing stock. This means there is a need to provide affordable housing in these areas as the less well off are priced out of the housing market.

Cultural factors, whilst mainly concerned with working overseas, can influence customer choices in the home country. The UK comprises a diverse population of different races and religions, from which there are many potential customers both in business and as individuals. Their ways of life and beliefs have to be understood to develop good customer relationships. Overseas, additionally, there are cultural practices which determine the means of carrying out business. A good example is the Chinese virtue of 'saving face', a very complex subject, but crucial to understand when working in this environment. In some countries women are not afforded equal legal status with men, which can have implications in staffing operations there. Working overseas can require different approaches to marketing, sometimes with moral implications as in parts of the world it would be expected to 'grease the palm' to obtain work or be offered contacts.

Technology development moves on apace and an understanding of the likely impact and opportunities is important in the marketing process, especially when involved in the design process, as accommodation for the future needs to be designed for and built in. A demonstration of the company's awareness and understanding of these issues could assist in obtaining work. The development of information technology and the accompanying communications hardware and software, increasingly relies on smaller and portable hardware and wireless technologies. Employees can now take their office with them, work from home or hot desk, affecting the type of office accommodation required. Whereas once there was one television set in the lounge, now small children have one in their room, along with computers and other gadgetry. There are many more labour-saving pieces of equipment all of which require servicing. How long before the provision for robotics occurs in the office and the home?

Lifestyle is continually altering as culture and prosperity change, technology being one of the main drivers. The use of the Internet has changed shopping habits, which affects the retail markets' building requirements. People are choosy when they look for their home; near good transport links and other facilities in the area.

Environment is now centre stage politically. Since buildings are significant contributors to carbon emissions both in terms of embodied energy and that required for heating, cooling and lighting, there is potentially a great marketing opportunity for companies involved in both the design and construction processes. Public knowledge, concern and perception have changed considerably over the last decade. New designs, innovative use of materials, those currently available and those in development, will be used. Methods of becoming more sustainable and self-reliant will be on the agenda, especially in terms of energy use. How lifestyles will change as a result is up for debate at the time of writing, but the industry needs to be ready to meet this demand and play its part in marketing the available solutions in the home market and for industry and commerce.

5.5 Forecasting

Forecasting is about reducing uncertainty. The more accurate the information is, the better the decisions are likely to be. It is also about looking at historical data to establish future trends. Accurate forecasting permits the business to plan for the future and gear up to have the appropriate resources in place to carry out the plan, and it can establish opportunities and likely threats in achieving the plan. Forecasting can be applied for the short, medium and long term.

Short-term forecasting in construction is most prominent in the production of the detailed programme of work for the contract. This is where a month or three months of the contract programme is produced in more detail to assist in controlling progress of the project.

Medium-term forecasting is the production of the master programme for the contract and the annual business plan, which is the control document for the business as a whole for the forthcoming year. This looks at the projected turnover necessary to cover the business overheads and reach the profit required. Both documents are used as the main control documents by monitoring actual performance against the forecast.

Long-term forecasting encompasses the strategic requirements of the business and can cover a period of up to five years. The company is deciding, from the position it is in now, where it wants or expects to be in five years' time taking account of the issues likely to hinder or assist this goal (Chapter 4). Once the strategic direction is decided, steps can be put in place to achieve this by developing the resources and advising the employees and shareholders of the intentions. Employees perform to their best when they are informed about their future so they can engage in its achievement, and shareholders want to know their investments are being protected.

5.6 Forecasting techniques

5.6.1 Expert opinion

Whether it is a small one-person business or a large organisation, expertise is an essential part of forecasting. The individual and the heads of sections have accumulated knowledge of what is realistic based upon their experience of the past, more finely tuned if they have collected data upon which to base their opinion. An experienced project manager can look at a set of drawings and determine how long it should take to erect the structural frame using differing techniques and so on. With data the planner can then confirm it is possible. Purchasing managers can predict the likely costs of building materials and sub-contractors if the contract is awarded. Readers of this text can budget their own personal finances based upon experience and the known cost outlays for the year. How successful they are depends on the level of optimism or pessimism they apply to the figures. Some of the knowledge can be sought from current and potential customers.

5.6.2 The jury method

Sometimes referred to as the expert panel method, this is a well-established means of forecasting. A panel of experts is brought together to proffer their forecasts and from this an agreed forecast is reached. An example of this is the monthly meeting of senior economists advising the governor of the Bank of England where interest rates should be set to control inflation. The success of this method is dependent upon the quality of members' expertise.

Meeting monthly gives the opportunity for their views to be amended than less frequent meetings do, which can cause a higher degree of uncertainty.

5.6.3 The Delphi method

This is also a panel of experts, but in this case they are not brought together under one roof so they cannot influence each other or follow the majority, if they feel it prudent to do so. A questionnaire is sent to each member of the panel asking for their predictions on key issues. The collected replies are then circulated to all members along with further more directed questions. This can take several rounds with the questions becoming more specific, until a detailed view is obtained.

5.6.4 Decision tree analysis

This type of forecasting looks at all the possible outcomes to the major factors affecting the business and then predicts the likely result. This can involve the use of statistical analysis. For example, the decision may centre on whether or not to diversify into civil engineering work. The key factors could be level of management experience needed, amount of projected work in the firm's area of activity, the level of competition, availability of experienced sub-contractors and so on. Taking the example of management experience, the tree would develop the questions: is it available in-house, and if so what retraining is required; can this be done in-house or not, and if not where can it be provided; what will it cost and how long will it take? If the experience is not available in-house, example questions would be: where is it obtained, how long will it take to recruit and what will it cost? From this and other branches of the tree information comes to light to inform the decision-making process.

5.7 Market positioning

So, what is the construction market? The industry used to divide itself into building and civil engineering, and those companies which did both, had two separate sections or companies. Perhaps this was at a time when there were clearly identifiable civil engineering projects such as the motorway programme. Building projects were smaller than they are today. In recent years in the UK, civil engineering projects are fewer and larger building projects are being carried out with a large civil engineering content; the two areas have merged out of necessity. This is significant, as the two marketing functions have also merged. The reality is that the industry covers a wide range of activities. In terms of type of construction, there is speculative housing, housing, hospitals, offices, educational establishments, factories, supermarkets, out-of-town shopping complexes, theatres, sports stadiums and facilities, fairgrounds and leisure facilities, prisons, petrol stations, fast food chains, airports, docks, roads and motorways, railways, docks and harbours, sea defences and bridges. Taking the speculative housing market alone, there are multi-storey flats, low-rise flats, maisonettes, executive, terraced, semi-detached and detached.

In terms of types of contracts, there are competitive, negotiated and serial tendering, design and build, construction management, management contracting, partnering agreements, etc. There are also maintenance and alteration contracts and, recently, facilities management contracts. Some are small and others worth several hundred million pounds. The customers are also wide-ranging, as discussed earlier.

So, the company must ask: which part of the market does the company wish to participate in and what penetration of this market does it wish to accomplish? This is a strategic senior management decision which analyses resources and the potential market. In essence, though, it covers three main areas: the type of product or service provided which also includes the type of customer; the geographical are the work is to be carried out in; and, especially in the case of the speculative market, the price range and quality standards.

5.8 New products or services

In construction, generally, a service is provided rather than a product, the obvious exceptions being the speculative housing market or the speculative commercial/industrial unit market. This is because the industry constructs a building the client has determined is required.

The first stage is to conduct a gap analysis. The market research may have indicated there is a gap in the market as a result of shifting requirements. An example of this is the changing composition of the population with the increase of one-parent families and the elderly population. Another is the massive increase in the numbers of people requiring accommodation in the centre of large cities. Offices and warehouses have been refurbished and new apartments have been constructed. Associated with this is the social infrastructure needed to support this change in the needs of the population in the inner cities, such as bars, clubs and restaurants. In the 1990s there was a major expansion in the out-of-town shopping complexes such as Trafford Centre in Manchester, Meadow Hall in Sheffield and Bluewater in Kent.

Planning restrictions have now brought out-of-town locations to a halt, and the emphasis has changed to inner-city developments such as the Bull Ring in Birmingham. Facilities management, a significant development often linked to PFI contracts is another example of a gap offering opportunities.

It is occasionally possible to create an opportunity. Many manufacturers are housed in out-of-date buildings that have been modified over the years to house the new plant and equipment necessary to be competitive. The buildings are often unsuitable for purpose and not energy efficient, but their greatest asset can be the land upon which they are constructed. There are examples where the developer has offered to construct a purpose made building to the latest building regulations, but on a new site, provided the owner gives them the land currently in use, either as full payment or in part exchange, the developer realising the profit they can make from developing the land for a different purpose.

Once a gap in the market has been identified it is necessary to establish what the likely competition is. It is interesting to note that nearly all the large, out-of-town shopping centres were built by Bovis as the managing contractors, primarily because the chief executive, Sir Frank Lampl, spotted the opportunity long before anyone else and put steps in place to ensure the company was positioned to procure the work: an example of vision. If there is competition, then the costs of resourcing the work needs to be analysed so a decision can be made whether or not there will be an acceptable contribution to the profit of the company.

5.9 Pricing decisions

This is a senior management decision. In the case of the tendering market this decision will be made at the time of tender, when senior management reviews the current state of the market and the business, and makes a decision on what percentage to add to the estimate. When the tender is being put together most bill items will be priced similarly by each contractor. Indeed, it is the estimator's job to establish what the true cost of building will be to enable senior management to assess the company's financial position in the marketplace and decide what profit margin and/or contribution to overheads the bid should include. There is also some flexibility in pricing the items by applying commercial judgement as to what the actual price will be if the tender is successful. In other words, the supplier or sub-contractor might be prepared to change the price they gave if the contract is awarded.

The speculative market is different as the building can only be sold if the price is right. The developer will have done its valuation calculations, such as the residual method or discounted cash flow analysis (*Finance and Control for Construction,* Chapter 3) to establish whether the figures will give the required profit level. There will also be some flexibility depending on the market at the time of sale. In the case of the speculative market the sums will already have been done prior to the project having been started.

Setting a price is a complex business and there are many influences that affect it. Besides wishing to make a profit, a buyer has to be found who is prepared to purchase at or near the required price. Speculative builders, anxious to ensure a sale, will often offer a fixed price, providing the purchaser pays a substantial deposit. This means the builder has a guaranteed sale and the purchaser will not experience any increases in price. Alternatively, they will ask for a small deposit but expect the purchaser to accept any increases in costs, the downside being if the rises are too much the purchaser may pull out of the contract. On a large retail development, the developer may offer a lower rental price to those that buy-in before the development is completed. All of this is set against a background of supply and demand that at times can be volatile. World events can affect the global and hence the country's economy. The introduction of home information packs in 2007 caused many sellers to put their homes on the market ahead of this to avoid the costs of the pack. The result was a slow down in house inflation, which affected the speculative market as well.

5.10 Promotion and selling

With exception of speculative builders and developers, construction companies do not have a brand to sell as do those involved in consumer products. What they do have is a company image to promote. Those images most relevant to the industry are to do with completion on time and on budget, to a prescribed level of quality, rapid reaction time to defects, a good safety record, paying suppliers and sub-contractors on time, not hiding behind the small print of the contract to make claims, and above all, looking after the client's and other third parties' interests. There a variety of places this message can be promoted, including television, radio, newspapers, the Internet and other electronic means, in-house publications, signboards, booklets, leaflets and word of mouth. Methods of promotions fall into four main categories: personnel selling, advertising, sales promotion and public relations.

Personnel selling in construction is used when making presentations to prospective clients, which can occur when tendering for a project and when applying to become a member of preferred lists. It occurs indirectly when attending conferences, seminars, exhibitions, professional institute functions, sponsored sporting and corporate events, which prospective and existing

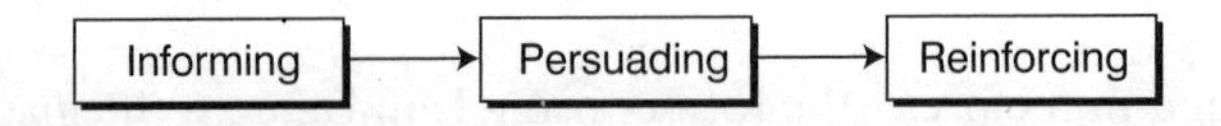

Figure 5.4 Setting the objectives

clients may also be attending. These can be important occasions and should not be seen as a 'junket'. They give the opportunity for businesses to meet people, nurture relationships and discover what is likely to be happening in the future. These long-term relationships develop feelings of trust and understanding and enhance the possibilities of obtaining work in the future.

In the speculative market selling normally occurs on the site, in offices in or adjacent to the show house or houses, depending on the scale of the development. Here staff, not always construction personnel, are on hand to show prospective customers around the premises and to provide information about the costs, types and availability of property, and choices available for the interior finishes and equipment. They may also be offering financial packages to aid purchasing and be able to discuss the amount of deposit required if the customer is interested.

Large developments, where the infrastructure is in place, such as Salford Quays or London's Docklands will also have a sales office on site, but their function will tend to be more commercial because of the nature of prospective clients, who may also be developers.

The first stage of advertising is setting the objectives, which is comprised of three progressive tasks as demonstrated in Figure 5.4.

Informing is telling prospective customers and the general public that the company is in business to provide the service of construction. Persuading is about creating an attitude in the 'informed' that the product or service being provided is one they want. Reinforcing is ensuring that once the customer has changed to using the company, they will wish to continue and as a result further business will follow. It does this with a message, which comprises five progressive stages as demonstrated in Figure 5.5.

The first stage is to create awareness ar 'attention getting'. This can be achieved with words or with pictures such as a logo. After World War II, the name Wimpey was synonymous with construction, most house buyers will have heard of Barrett, and Taylor Woodrow has used their logo of four men pulling together with great effect. Both Wimpey and Barrett once spent a lot of time and effort promoting their house-building programme on both

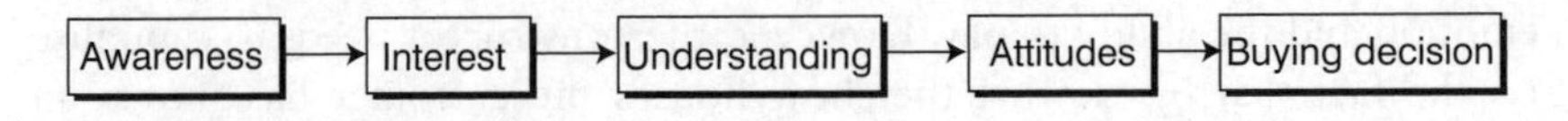

Figure 5.5 Creating the message (reproduced from Czinkota *et al.* (1997) with permission of Wiley-Blackwell).

television and in the press, the former using Paul Daniels, the magician, and latter the actor Patrick Allen flying a helicopter.

Once the audience is attentive, the next stage is to create interest so the reader or viewer continues to want further information. An example of this is the headlines shown on the board outside the newsagent advertising the evening edition of the local paper such as 'local hero rewarded'. Television and cinema use trailers of a film or programme they wish to encourage people to see. The construction industry sometimes uses artists' impressions of completed buildings.

Once interested, there needs to be information made available so the prospective client understands what they are going to get for their money; that the company is equipped and able to carry out projects to the client's satisfaction and standards, and that the project is going to suit their needs. In speculative housing this will usually mean providing a show house or houses to demonstrate what the final building will look like, as the majority of purchasers will have difficulty in relating two-dimensional drawings of the building to the final three-dimensional form.

The next stage is to persuade the prospective client that the product or service provided is what they need. The most successful show houses are those fully furnished in the style suiting the lifestyle of the prospective customers. Small inner-city flats for young couples will be different from expensive four- and five-bedroom detached houses. Placing appropriate toys in the child's bedroom and allowing them to play with them whilst parents view the rest of the house is trying to create an attitude of desire, as does a well-equipped kitchen to whoever does the cooking. Cleverly designed landscaping gives a feel of what it will be like on a nice summer day in the garden. Similarly, a high-quality presentation to a client by a construction company is done to create a positive attitude. Some companies now use a video recording of the key personnel as part of the presentation, rather than allowing them to perform live. This means the personnel can get it right before the presentation and can be filmed with different backgrounds, such as on sites of successful previous projects. It is not possible to be specific here as the nature and style of presentations is continually changing as are clients' expectations.

Finally, the buying decision has to be made by the client. In the case of the speculative housing market, this can be quite rapid, much based upon emotion and the ability to pay. However, the constructor needs to capitalise on the fact that by viewing the show houses the customer has shown an interest. Offering other services, such as financial and legal advice, discussing alternative finishes, and keeping in touch with them by mail or phone reinforces the interest shown. It should be noted that this has to be done carefully as over zealous contact can have a negative effect.

5.11 Choosing the media

The majority of advertising is carried out in the press and on television. Less frequently used are radio, billboards and cinema. Television is at its best when showing action, so showing a pictures of the house and its interior is not a very stimulating experience for the viewer. Advertising space on television is also expensive, so to have an advert showing a moving view through the property would be very expensive because of the time needed to do it justice. Hence the use of personalities to talk about the product so there is an association with the celebrity and the building. Another approach is to use still photos to show where the site is, with words, both on the screen and spoken, to excite the viewer to go and have a look. This becomes a shorter advertisement and is therefore cheaper. However the costs are still high and generally are used by only the larger builders. Unlike a newspaper advertisement, the viewer cannot refer back to it, so it has to be shown repeatedly over a short period of time to obtain the reinforcement mentioned earlier. Where television can be more profitably used is to promote the company as a by-product of a news item such as the opening of a prestigious building.

The press comprises national newspapers, regional newspapers, magazines and technical and trade publications. The national press is basically divided into two categories, the tabloids and the broadsheets, each with it own demographic of readers. Usually at the weekends they produce a colour supplement with featured articles rather than news items. These rely heavily on revenue from advertising. Regional papers are either daily or weekly; some have readerships covering large rural areas, whilst others cover densely populated areas such large towns and cities and others may be aimed at a smaller towns and communities. Magazines cover every range of the public's interests, but tend to be very specific by nature, dividing the reading audience by age, gender and interests. Trade and technical papers are specific to an industry, trade or supplier group. Where the contractor believes the audience to be will determine where it places an advertisement. Again, as with television, there will be opportunities to publicise the company as part of the ongoing public relations strategy.

Whilst senior management may have a view on the composition and method of presentation, it is a very specialist business with its own expertise, and advice should be sought if it is not available in-house. If the company goes to an agency with a predetermined view as to how the advertising campaign should go, then of course the agency will quote and produce whatever is required. However, if the company goes with an open mind expressing the overall aim of the campaign and leaves the agency to develop ideas freely, a much more successful outcome is likely.

5.12 Public relations

'Public relations (PR) is being good and getting credit for it' is an important adage and it is a misconception that PR is about successfully promoting, irrespective of quality. It has to be of sufficient consistency and quality to engender interest. There will be incidents when a PR exercise will be conducted to limit damages when something has gone wrong such as a serious accident, but this only is effective if it is an occasional event rather than regular and, in any case, is more about controlling the outward flow of information from the company, especially to the media.

Every member of the company who talks to or meets people from outside is indirectly in the process of PR. Their attitude and manner conveys signals as to the type of organisation they are representing. Indeed, every time an employee picks up the telephone to answer an outside call, a PR exercise takes place. The manner and way the employee reacts is important. One of the most important people in this chain is the receptionist, as this is often the first person to whom the client speaks.

The role of PR is to be the interface between the company and its employees, and people and organisations outside the business; Figure 5.6 demonstrates the wide range. PR has several objectives, but in general it is to foster the reputation of the company, to promote the services offered, and in the case of the speculative market, the buildings.

High on the priority list is communication and liaison with the media. The way this is carried out is determined by the overall marketing strategy and the amount of finance available for it, which in turn will dictate which section of the media is primarily to be used. At one end of the scale, large amounts of money can be used for producing advertisements for broadcasting on television, whilst at the other are newsworthy stories the media will use at no cost to the organisation other than the development and presentation of the information. The media will only use factual stories and if they find the story given is not truthful, the reputation of the PR department will suffer and they may find difficulty in having stories accepted in the future. The problem with any story given to the press is that once handed over, the company loses control on what is actually written or broadcast. It is standard practice to produce press releases with a time embargo, which means the media cannot publish until the date or time given. These can be given to the media and they can also be supported with

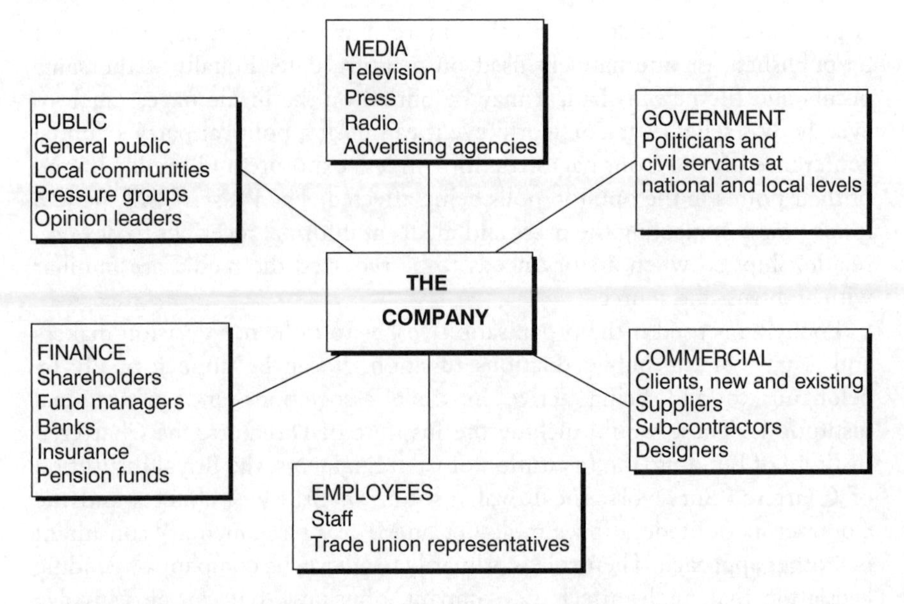

Figure 5.6 The PR interface (adapted from Jobber (1995))

a press conference if the story is significant. The press release will be used in this case as encouragement for the press to attend. Press conferences are usually called if the news to be announced is significant, such as a merger, takeover of a major company, the appointment of a significant personality, or resignation.

Writing press releases is a skill, but detailed below are a few key issues that should be addressed. Remember: the media receiving the release needs to be convinced it is a story worth publishing. There should be a headline that grasps the reader's attention that is factual and to the point such as, 'The Prince of Wales will officially open the new offices of Winston Construction' or 'Winston Construction will pay workers at least £50,000 per annum to work on the new shopping centre at Newtown'. The opening paragraph should be a brief summary of the content of the release, followed by the key points in descending order of importance and, as with the headline, should be factual and to the point. The overall length should be as short as possible, ideally no longer than one page. The editor of the newspaper or programme is busy and needs to be excited as quickly as possible otherwise the document will very rapidly find its way to the wastepaper basket.

Timing is important for the greatest impact, but cannot be guaranteed. The editors will publish information when there is room for it and this is a function of its relevant importance to other news that breaks on the same day. This can have advantages and disadvantages. If the company's news is good and it coincides with other more newsworthy items, it may not be published, or alternatively used on a future date. Equally if the same occurs and the news is bad, it may be buried on the inside pages much to everybody's relief. A major world event can push a political party's annual conference off the front pages resulting in less exposure and the likelihood of their points in the opinion polls being affected. The PR staff will spend a lot of time just meeting the press and media at informal meetings to develop relationships so when a story needs to be reported the media are familiar with and trust the source.

Lobbying is used if the organisation wishes to influence decision makers and is part of the public relations function. It can be done formally by belonging to and being active in trade associations and professional institutions. These would include the Institute of Directors, the Chartered Institute of Building, the Institution of Civil Engineers, the Royal Institution of Chartered Surveyors, the Royal Institute of British Architects and the Contractors Confederation. Employing an MP as a parliamentary consultant is another approach. Their role is primarily to advise the company of pending legislation that might affect the company's business, but can also involve making contact with the appropriate minister. Informal routes involve getting to know MPs, local councillors, members of the chamber of commerce and resident associations.

The sponsorship of sporting events or the arts is a means of associating the companies name with a well-established event or team and, although also advertising the company, demonstrates the company's interest in the wider community. This can be done at national and international levels. Good causes are also areas of consideration, such as sponsoring charity walks, runs and fund raising events. It can be as simple as providing a lorry for the local carnival parade to participating in national campaigns for fighting cancer.

Having a presence at trade shows can be a means of bringing the company to the attention of others, especially if involved in a specialist market or to demonstrate credentials in a specific area of expertise. This is relevant to house builders who may wish, for example, to have a presence at the Ideal Homes exhibition.

PR is also concerned with controlling information if the news is potentially harmful to the business such as a serious accident or protest. It is normal practice when an accident takes place to instruct all staff not to talk to the press, but direct them to a named person, who will deal with the situation. Reporters will phone the site seeking information and these can be readily transferred to the person charged with communicating with the press, but they will also stand outside the site and approach personnel as they leave. Even when the accident occurs on a different site to a different company, it can have an impact. For example, the author was once instructed not to talk to the press, but to refer them to the PR department, after the progressive collapse of the multi-storey flats at Ronan Point. Another company using a different system constructed these, however the joint design between the floors and the wall was similar to that used by the author's company, the one at Ronan Point being an earlier version. Protests at the construction of buildings are either because of the inconvenience caused to locals or due to pressure groups not wanting the project to have started in the first place. In the former case, much can be anticipated and steps taken in advance, not just to reduce the inconvenience, but also to contact and liaise with local residents and those most affected, before and during the construction process. The latter is more difficult, as the protesters are already in place and will obtain high levels of publicity when the contractor starts on site. The problem for the contractor, besides losing time, is that the press see the people dealing with the situation adorned with the company's name as well as the site signage.

It is not always possible to control information and the author, on a job interview, arriving early went for lunch to the local hostelry which was full of striking workers, overhearing the conversation was quite enlightening. Any event the press are invited to needs to be stage managed. It is essential that all safety issues are covered, especially if the event necessitates walking on the site itself. Signs should be strategically placed, and cameramen advised of them and other interesting things, because when a film is made for a television report, the director needs 'link' footage to move from one issue to another if there is a break in the filming. All personnel should be wearing safety helmets with the company logo.

Internally, PR can be applied by informing the staff about the company. This is normally accomplished by using an in-house magazine or leaflets. Staff are interested in the future prospects of the company, so new contracts awarded and their value demonstrates what future prospects are. Rumours also occur within any organisation, especially about reorganisation, mergers and takeovers. The magazine is an ideal place to put the record straight, although this can also be done using internal memoranda. There is always an interest in knowing who has been promoted and who has left the company. Staff profiles can be used, for example a senior person joining the company or an achievement such as a sporting medal, receiving an honour from the queen or raising money for charity. Items on new technology used on a site is of interest as are high-profile opening ceremonies of the completed project. All of these are aimed at making the reader feel included in the organisation.

Finally, the amount of exposure of key personnel needs to be controlled. On the surface it seems odd that there can be too much exposure, however, after a time the public can lose interest in the message or can become bored or irritated by seeing the same person over and over again.

References

Czinkota, M.R., Kotabe, M. and Mercer, D. (1997) *Marketing Management: Text & Cases*. Blackwell Business.
Hague, P. (1993) *Questionnaire Design*. Kogan Page.
Jobber, D. (1995) *Principles and Practice of Marketing*. McGraw-Hill.
Mercer, D. (1996) *Marketing*, 2nd edn. Blackwell Business.
Stokes, D. (1997) *Marketing: A Case Study Approach*, 2nd edn. Letts Educational.

CHAPTER

Leadership and teambuilding

6.1 Introduction

Below are two descriptions on leadership. The first is to demonstrate that it is very difficult to become a great leader, and the second that there is not a definitive definition of a leader and much depends on one's own personality and talents.

> Dear Lord, help me to become the kind of leader my management would like me to be. Give me the mysterious something which will enable me at all times satisfactorily to explain policies, rules, regulations and procedures to my workers even when they have never been explained to me. Help me teach and to train the uninterested and dim-witted without ever losing my patience or my temper. Give me that love for my fellow man which passeth all understanding so that I may lead the recalcitrant, obstinate, no-good worker into the paths of righteousness by my own example, and by soft persuading remonstrance, instead of busting him on the nose. Instil into my inner-being tranquillity and peace of mind that no longer will I wake from my restless sleep in the middle of the night crying out 'what has the boss got that I haven't got and how did he get it?' Teach me to smile if it kills me. Make me a better leader of men by helping develop larger and greater qualities of understanding, tolerance, sympathy, wisdom, perspective, equanimity, mind reading and second sight. And when, Dear Lord, Thou has helped me to achieve the high pinnacle my management has prescribed for me and when I have become the paragon of all supervisory virtues in this earthly world, Dear Lord, move over. Amen.
>
> 'A Leader's Prayer', Charles Handy, *Understanding Organisations* (© Charles Handy, 1976, 1981, 1985; courtesy of Penguin Books)

Some advice given by Mike Stoney, Director of Laing PLC, to construction management students:

> Don't try to be a specific type of leader. All people are different and leadership style is a personal thing. Observe leaders, both good and bad, and pick out the good things that you think might suit you in your own development, and learn from the mistakes others have made.

6.2 Definitions of leadership

Various definitions have been written over the years, but in essence the five given below perhaps summarise the essence of the subject.

1. Leadership is the process of influencing individual and group activities towards achieving the objective.
2. The organisation, the leader and the group may have different objectives and interests. The leader's role is to find the right balance between these.
3. A leader is one who succeeds in making others in the group follow their lead.
4. A good leader enthuses others in the group to want to achieve the goal.
5. They are the ones who create the environment that others are motivated to work in.

It is a misconception to think that managers are the leaders. They may well be, but not necessarily. A manager has responsibility, and therefore occupies a formal position in the organisation, but does not always have the ability to lead others as defined above. It is not uncommon to find a person below the manager is the one who leads and actually gets things done.

6.3 The power of leadership

Leaders require power to be able to lead, but power comes from a variety of sources. These include:

- Legitimate power that results from the position one holds within the hierarchy of the organisation and the authority that goes with that position.
- Reward power goes with the ability to be able to reward people for their work. This can include financial rewards, such as pay rises, bonuses

and promotion, but also non-financial rewards such as recognition, staff development programmes, interesting or high-profile work.

- Coercive power is the ability to punish subordinates in a variety of ways. These include reprimands, criticisms, suspensions, warning letters, demotions and termination of employment. This should not be confused with bullying.
- Information power is when a manager has access and control over information that subordinates do not have. This can also work the opposite direction when a subordinate has knowledge and access the manager does not have, such as IT skills. This overlaps with the next item.
- Expert power is expertise accumulated and valued by others because the subordinate's expectation is to learn about them and assist in their future promotion and success. This, as with information power, can work in reverse, for example, on site many of the support functions such as planning and quantity surveying are specialist. If the manager comes from a different background he can be 'subservient' to those with this expertise.
- Personality power is when a manger is respected and admired, but not necessarily liked. Some of our great leaders also have charisma, something difficult to quantify, but they tend to 'fill the room' when they are present.

6.4 Leadership style and types

There are various ways leadership style is defined, some more complex then others. Those given below are only a summary of some of these and are based on the behaviour of the leader rather than a measurement of the skills of leadership.

6.4.1 Autocratic

These people believe that decisions and the authority to make them must remain with the leader. This is often because they believe that the subordinates are incompetent and lazy, reflecting McGregor's Theory X (Chapter 1). When they give orders the subordinates are expected to follow.

The advantage of this style is that tasks are completed efficiently as there is no time for two-way communication, providing the leader is competent. The problem is subordinates are told what to do and not why. They follow instructions even knowing they are wrong. This type of leadership normally leads to poor morale and low productivity.

6.4.2 Democratic or participative

These leaders delegate authority to subordinates and allow them to make some decisions depending upon their perceived competence and interest in dealing with the task. The leader involves the subordinates in discussing the objectives of the task, developing strategies to accomplish it and determining roles.

This style of leadership improves productivity because subordinates are engaged in the decision making and feel useful and have increased job satisfaction as a result. Because they are involved in the decision-making process, decisions made are of a higher quality because of the extra thought given, and if it is necessary to change the methods of work or control mechanisms they are less resistant to it because they have an understanding of why change is required.

However, this assumes the subordinates wish to be involved in participation, which is not always the case. Whether or not this is due to the way management conducts themselves can be a factor. However, it is clear that if management is just consulting rather than fully involving them, subordinates will soon lose interest and stop co-operating in the process. Often subordinates have been conditioned to do as they are told and not be consulted. To be confronted with participation is sometimes a culture shock they may not wish to become involved in.

6.4.3 Laissez-faire

Here the leader abdicates the leadership position. They handle the group loosely allowing them to do more or less what they want, usually handing over the leadership to someone else in the group. The usual reasons for this stem from lack of confidence, fear of failure or wanting to be part of the group, especially they were promoted from within it. It can work if the group is very experienced and highly motivated, but it is more likely to fail.

A further way to classify leaders is as Likert suggested: in the manner in which they get the task done. He suggested there were two types: task or production oriented or people- or employee-centred leaders (Chapter 1).

6.5 Theories of leadership

There are several theories of leadership which have developed over the years, the earlier ones based on observation and assumption rather than scientific study.

6.5.1 The great man approach

The adjective great should not be confused with moral. Indeed many great leaders in history have been immoral. The great man approach is based on the assumption that men and women of great vision, personality and ability rise to positions of prominence and affect and change the course of history. There is an underlying assumption that great leaders are born and not made. Well-known historical leaders who are used to demonstrate the great man approach are: Alexander the Great, Genghis Khan, Julius Caesar, Joan of Arc, Elizabeth I, Napoleon Bonaparte, Isambard Kingdom Brunel, Winston Churchill, Mahatma Gandhi, Joseph Stalin, Mao Tse Tung, Henry Ford, Martin Luther King, Nelson Mandela and Bill Gates.

Advocates of this approach also believe that very successful or great men and women can be found in certain families. In modern times there are examples of dynasties in certain areas of life, such as in politics with the Kennedys and the Bushs, and in theatre with the Redgraves.

The problem with this approach is that if it accepted leadership is inherited, favouritism in promotions is inevitable. If it is believed that leaders are born and not made, then all leaders in the organisation have to be staffed with born leaders. This will inevitably cause a problem because it is clear from the examples of great leaders, there will be insufficient to go around. Further, if leaders are born, then there is no point in providing staff development for others who were not born to lead. Finally, since leaders are born, they will define their own parameters for the job and the concept of producing job roles and specifications becomes redundant. It can be seen from this argument that whilst it is true to say there are occasionally 'great' born leaders, it can be concluded this approach is severely limited.

6.5.2 The trait approach

Similar to the great man theory, this case assumes a leader's personal attributes are the key to leadership. Traits are distinctive qualities or characteristics an individual possesses and can include physical attributes, personality, skills and abilities, and personable skills, such as the ability to socially interact with others.

The difference with the great man theory is that the trait theorists do not necessarily assume leaders are born with these traits, but only need some of them. Examples of suggested traits have included height and stature, vitality and energy, drive, enthusiasm, decisiveness, good looks, knowledge and intelligence, imagination, sociability and friendliness, courage, self-confidence, honesty and integrity, fluency of expression, and self-control.

Researchers have found that several traits seem to be related to effective leadership. Ralph Stogdill (1974) found that certain characteristics seem to be important to effective leadership: The traits include:

- being able to adapt to the situation;
- being alert and aware of the social environment in which one is working.
- being achievement oriented;
- being assertive, co-operative, decisive, persistent, self-confident and dependable;
- having the desire to influence people;
- having high levels of energy and a tolerance to stress;
- being prepared to take responsibility.

The skills include:

- being intelligent and having common sense;
- being creative and conceptually skilled;
- being tactful and diplomatic when required;
- being persuasive and socially skilled;
- having organisational and administrative ability;
- having good control of language;
- having full knowledge of the group task and required outcomes.

Edwin Ghiselli (1971) found the main characteristics for effective leadership were:

- supervisory ability or performing basic functions of management, especially leading and controlling the work of others;
- need for occupational achievement including seeking responsibility and desiring success;
- intelligence such as the quality of judgement, reasoning and reactive thinking;
- decisiveness in having the ability to make decisions and solve problems capably and well;
- self-assurance in believing oneself capable of coping with problems;
- initiative in having the ability to act independently, develop courses of action not readily apparent to others and to find new or innovatory ways to do things.

Whilst on the surface this approach seems to make sense and there is much validity, there are certain limitations to the trait theories. Just because one has some of these traits, does not necessarily make a person a good leader. There are many tall people who fail and historically many small people who have made 'great' leaders such as Napoleon and Ghandi. Many academics have intelligence and are very knowledgeable, but by no means all make good leaders.

The definition of the traits can be misunderstood and different interpretations made. For example, in the previous paragraph the word intelligence is used. Many will interpret this as being 'brainy' and educated, but there are many intelligent people who have not had the benefit of a good education, but who have a high level of commonsense and who are equally if not more intelligent than the former. Others would consider a great athlete to be 'intelligent' because of how their brain works to co-ordinate the limbs, reaction times and vision. How do you define dependability? Is it being trustworthy or always being there when needed?

It is often assumed a leader has good values, which in turn gains respect. Research in prisons has demonstrated that often the most effective leaders within the prison populations are homosexual, neurotic and psychopathic. Further, trait theories imply success is determined by inherent qualities and ignore the environment in which people have been brought up.

How can traits be measured? Height can be physically measured as can intelligence to a certain extent, but how to measure looks, which is very subjective and cultural. Some Eastern societies believe that tanned skin is bad as it implies field work, whereas many in the West strive to be tan. Even if it can be measured, what scale is used and can one trait be compared against another using the same scale? Some people in a given job will demonstrate traits they believe to be required even though it goes against their natural tendency. For example, they may be naturally gregarious, but try to remain aloof and distant from subordinates, or they make quick decisions rather than give them the thought they would normally. Finally, most people in society want a leader, so anybody who decides to become one may well be followed irrespective of ability. However, in spite of the points made above it does not mean trait theory is totally invalid.

6.5.3 The behavioural approach

The concept behind this is to concentrate on the behaviours of leaders which can be measured. Behaviourists assume leaders are not born, but can be developed through a learning process. Hence the focus is on what leaders do rather than what they are.

In the 1930s, Kurt Lewin (1890–1947) carried out research on small group behaviour, which had a great influence on further researchers. He divided a bunch of boys into small groups to which each was allocated an adult leader. One of these was autocratic, another democratic and the other laissez-faire (section 6.4.3). In the group lead by the autocratic leader there was no long-range planning, a considerable amount of aggression occurred and, whilst present, productivity was similar to the group led by the democratic leader, although the quality was lower. When the leader left, productivity stopped altogether. Two of the boys left the group because they believed they had been made scapegoats for the failure of the group. In the group led by the democratic leader, work continued even when the leader left. Productivity was lowest in the group led by the laissez-faire leader, the boys did as they pleased and they all became frustrated. In conclusion, it was seen the democratically led group was the most successful. To complete the experiment, the leaders were moved around to take charge of the other groups. In each case, after an initial adjustment in the boys getting used to the new style of leadership, the groups performed the same as in the first allocation of leaders.

This was followed by two major research projects at Ohio State University and the University of Michigan carried out in the 1940s. Ohio's aims were to identify the behaviours exhibited by leaders, to determine the effect these behaviours had on both employee satisfaction and performance, and to identify the best leadership style. By questioning both civilian and military personnel about the behaviour of their supervisors, they found subordinates generally saw their leadership behaviour in two categories that they called the 'initiating structure' and 'consideration'. Initiating structure is the extent to which a leader defines the activities of subordinates in attaining their organisational goals, sometimes referred to as being production- or task-oriented. Consideration is the extent to which leaders are concerned with developing mutual trust between themselves and their subordinates as well as respect for their feelings and opinions, sometimes referred to as being employee- or human-relations oriented. The Michigan studies' conclusions were very similar.

Other views on leadership are the managerial grid produced by Blake and Mouton, McGregor's Theories X and Y and Likert's systems of management as described in Chapter 1.

6.5.4 Contingency theories of leadership

Both the great man and trait theories are of limited use, and the behavioural approach did not produce the answers. This was mainly because on investigation of different successful leaders, whilst many did, not all fell into the categories defined above, so the contingency or situational approach was adopted. In other words, there is no leadership style appropriate to every manager to cover all circumstances.

The Fiedler model

Fred Fiedler developed one of the first and best known leadership contingency theories. The basic premise behind it is that the effectiveness of a group or an organisation depends on the interaction between the leader's attitude and personality, and the situation. The situation is defined as the extent the leader has the power, control and influence over the situation, and the amount of uncertainty the situation causes the leader. This clearly differs depending on the type of situation. A soldier in battle has a different expectation of the leader than a secretary in the office. Fiedler's contingency model is used to identify which type of leader will probably do best. His least preferred co-worker scale (LPC) as shown in Table 6.1, reproduced with permission of Professor Fiedler, identifies 16 personality traits which the leader considers best describes the person(s) for whom he or she is responsible. Generally, if the leader describes the others in the team negatively, i.e. a low score, they tend to be task oriented, and if the tendency is to describe them positively, i.e. a high score, they are more likely to be people oriented.

After the person's basic leadership style has been assessed using the LPC scale, the situation is evaluated to match the leader's style with the specific

Table 6.1 Traits used by Fiedler to establish the least preferred co-worker (LPC) scale

Pleasant	8	7	6	5	4	3	2	1	Unpleasant
Friendly	8	7	6	5	4	3	2	1	Unfriendly
Rejecting	8	7	6	5	4	3	2	1	Accepting
Helpful	8	7	6	5	4	3	2	1	Frustrating
Unenthusiastic	8	7	6	5	4	3	2	1	Enthusiastic
Tense	8	7	6	5	4	3	2	1	Relaxed
Distant	8	7	6	5	4	3	2	1	Close
Cold	8	7	6	5	4	3	2	1	Warm
Co-operative	8	7	6	5	4	3	2	1	Unco-operative
Supportive	8	7	6	5	4	3	2	1	Hostile
Boring	8	7	6	5	4	3	2	1	Interesting
Quarrelsome	8	7	6	5	4	3	2	1	Harmonious
Self-assured	8	7	6	5	4	3	2	1	Hesitant
Efficient	8	7	6	5	4	3	2	1	Inefficient
Gloomy	8	7	6	5	4	3	2	1	Cheerful
Open	8	7	6	5	4	3	2	1	Guarded

situation. Fiedler argued there are three situational variables that interact with the leader's style which determine the effectiveness of a leader. These are:

1. *Leader–member relations.* This is the extent to which the subordinates support the leader. To the leader this is a measure of how loyal and trustworthy the subordinates are when given an instruction. Shown as good or poor in the model demonstrated in Figure 6.1.
2. *Leader position power.* This is the amount of power the leader has been given by the organisation to carry out the tasks. The leader may have the authority to dole out rewards or punishment, but if this has not been given, the leader relies on other methods to influence subordinates to obey instructions. Shown as strong or weak in the model.
3. *Task structure.* This is about the clarity of tasks; how clear are the directions given for each stage of the task? If the task is highly structured, with detailed directions, it is easier for the leader to monitor performance and influence the way the task is carried out by the subordinate. If the task is unstructured, the leader has not decided, or does not know the best way to carry out the task, the subordinate's performance cannot be monitored. Shown as high or low in the model

Fiedler concluded that task-oriented leaders generally perform better in situations which were favourable to them (categories 1–3 and 7–8) and less well in situations which were unfavourable. Relationship oriented leaders perform better in moderately favourable conditions (categories 4–6) as demonstrated in Figure 6.1. He believed the leadership style was fixed so the only way to improve the effectiveness of the leader was to change the situation. An analogy would be in cricket, where a bowler can only bowl to his capabilities and the batsman to his. Therefore the captain will change the bowler depending on the style of the batsman and the state of the pitch and the weather.

Leader-participation model or the normative leadership model

Developed by Victor Vroom and Philip Yetton in 1973, the leader-participation model relates leadership behaviour and participation to the decision-making process. It is based on the quality and acceptance of the leader's decision. Unlike the trait and great man theories, it is assumed leadership style can be adapted to suit the situation and is not necessarily inherent in the leader.

They defined a range of leadership styles, all of which they argued can be effective, depending on their answers to a set of questions on quality, information, structure, commitment, goals and conflict. The answers to

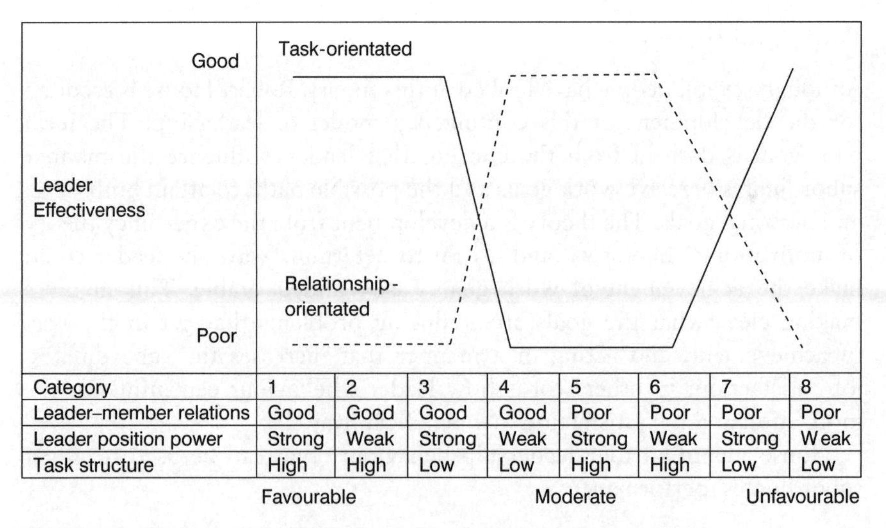

Category	1	2	3	4	5	6	7	8
Leader–member relations	Good	Good	Good	Good	Poor	Poor	Poor	Poor
Leader position power	Strong	Weak	Strong	Weak	Strong	Weak	Strong	Weak
Task structure	High	High	Low	Low	High	High	Low	Low

Figure 6.1 Findings of the Fiedler model (reproduced with permission)

these questions direct the leader to the style that will be the most effective. These are as adapted from Vroom and Yetton (1973):

1. You solve the problem or make the decision yourself, using information available to you at the time.
2. You obtain the necessary information from your subordinates, and then decide the solution to the problem yourself. You may or may not tell the subordinates what the problem is in getting the information from them. The role played by your subordinates in making the decision is clearly one of providing the necessary information to you, rather than generating or evaluating alternative solutions.
3. You share the problem with the relevant subordinates individually, getting their ideas and suggestions without bringing them together as a group. Then you make the decision, which may or may not reflect your subordinates' influence.
4. You share the problem with your subordinates as a group, obtaining their collective ideas and suggestions. Then you make the decision, which may or may not reflect your subordinates' influence.
5. You share the problem with you subordinates as a group. Together you generate and evaluate alternatives and attempt to reach agreement on a solution. Your role is more like that of a chairman. You do not try to influence the group to adopt 'your' solution, and you are willing to accept and implement any solution, which has the support of the entire group.

Path–goal theory

Although several people have looked at this theory, Robert House is credited for the development of this contingency model of leadership. The term path–goal is derived from the concept that leaders influence the manner subordinates perceive work goals and the possible paths to attain both work and personal goals. The theory is a development from the expectancy theory of motivation (Chapter 1), and uses it to determine ways the leader could make the achievement of work goals easier and desirable. This involves making clear what the goals are, reducing problems that get in the way of achievement, and acting in a manner that increases the subordinates' job satisfaction. In other words how leader's behaviour can influence the motivation and job satisfaction in a beneficial way.

House identified four leadership behaviours that can be used to affect subordinates' performance:

- Directive leadership is about letting subordinates know what is expected of them. This involves giving them specific instructions and guidance. The subordinates are then expected to follow the instructions to the letter. This is similar to the Ohio study's initiating structure or task-oriented type.
- Supportive leadership behaviour shows concern for the needs and welfare of the subordinates by being friendly and approachable. This is similar to the Ohio study's consideration or relationship-oriented type.
- Participative leadership behaviour is about consulting subordinates, encouraging them to participate and then evaluating their views and suggestions before making a decision.
- Achievement-oriented leadership behaviour sets challenging goals for subordinates. The aim is to improve their performance and demonstrate confidence in their ability to complete the tasks well.

The path–goal theory argues the behaviour of the leader in one situation should not necessarily be the same as in another. It depends on the situation, and the leader should adapt his or her behaviour to suit the situation. This is demonstrated in Figure 6.2, which shows the behaviour of the leader as defined above, being influenced by the demands of the task and the characteristics of the subordinates, such as their experience and perceived ability, resulting in the outcomes of performance and satisfaction of the subordinates.

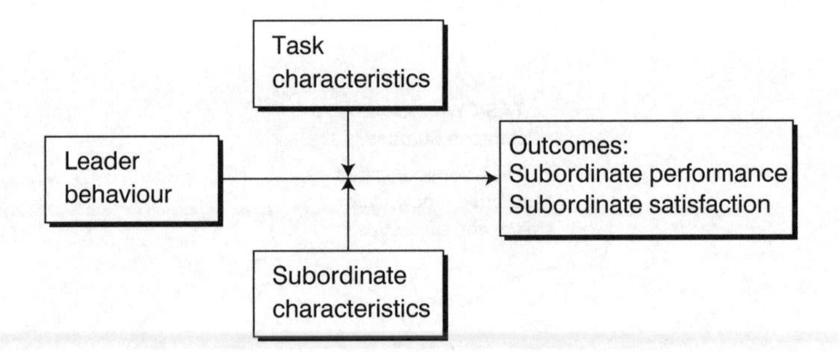

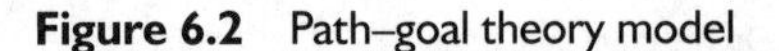
Figure 6.2 Path–goal theory model

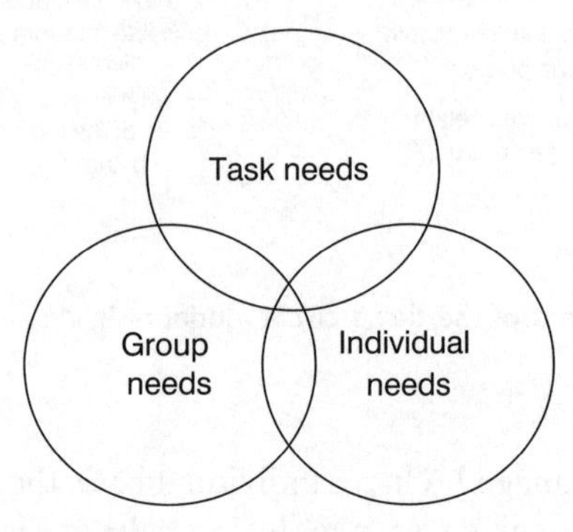

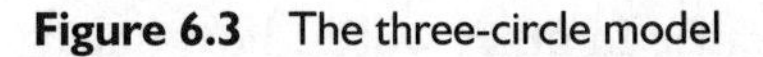
Figure 6.3 The three-circle model

Three-circle model of leadership

A further approach to leadership is that the leader of the team is related to the task to be carried out and the time available to complete it. It also considers the needs of the group of people making up the team and their individual needs, as shown Figure 6.3. The team leader has to understand all three aspects to make decisions and achieve the objective of the task.

Adair (1983) developed the three-circle model of leadership by proposing a number of questions the leader might ask, as shown in Figure 6.4.

The leader has to be clear about the objectives of the task and what has to be achieved. However, the rest of the team may not be clear and equally may disagree with the way this objective is to be met. The leader has to communicate clearly with all members of the team and engage them in the process, so they all pull together. The team leader has to understand the group composition and the various talents within so work can be allocated to these strengths and at the same time understand how each of the individuals within in the group relate to each other so conflicts between individuals can

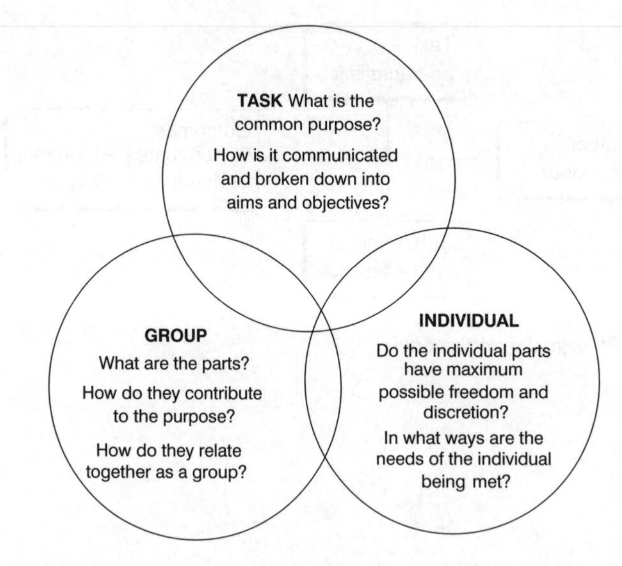

Figure 6.4 Development of the three-circle model of leadership (adapted from Adair, (1983))

be anticipated and managed. On an individual basis the leader has to decide how much latitude is given to people to make decisions for themselves with the best interests of the team at the centre. In other words, how much delegation of responsibility is given to meet the aspirations and ambitions of the individual.

6.6 Female and male leaders

Whilst perhaps controversial, an introductory chapter on leadership would not be complete by raising the issue on whether female and male leaders carry out their functions differently. Historically, leaders have been generally men with notable exceptions in England such as Boudica (Boadicea) and Elizabeth I, but until recently they have been relatively few. In modern times the world has had several eminent female politicians, such as Golda Meir (Israel), Sirimavo Bandaranaika (Sri Lanka), Indira Gandhi (India), Margaret Thatcher (UK), Gro Harlem Brundtland (Norway) and Angela Merkel (Germany). Increasingly, women are taking senior roles in industry and commerce, in what is still a male-dominated workplace, and it is anticipated this trend will continue, so the question posed at the onset is a valid and relevant one.

A word of caution as pointed out by Robbins (1994): if the styles of women and men are different, could this mean one or the other is inferior and if they are, is there a danger that gender leadership style can be stereotyped. Research appears to indicate the two sexes do use different styles. Women tend to use a more democratic and participative style by encouraging participation and sharing power and information. Men tend to be more direct by commanding and controlling using rewards for good performance and punishment for poor performance, although it should be noted not all do. However, it has also been found that when women are in male-dominated jobs they allow their natural styles to be over-ridden and they act more like men do in that position.

All the theories lean towards the needs for teamwork, information sharing, trust and the ability to listen and support subordinates as well as provide motivation. Generally, women do this better than men and have the ability to negotiate and compromise rather than try to win and make the other lose. It is interesting to refer back to Mary Parker Follet (Chapter 1) who was indicating this approach nearly one hundred years ago.

6.6 Further thoughts

The author has observed that better leaders generally show the following characteristics:

- They stand back, sometimes aloof, from the action and direct the talent in their control to resolve the problem, rather than mucking in, except when the extra pair hands will clearly be of use. This means that they can see what is happening which they would not see if they were actively engaged.
- They always seem to have time. Their diaries are rarely full giving them time to think, and appointments to see them can almost invariably be arranged quickly. Those managers whose diaries are full for the next month or more, generally do not make good leaders.
- They always appear to know what questions to ask that get straight to the heart of the problem. These challenge those present to be positive and imaginative. The question 'why not?' is a typical example and has floored many an expert who has given reasons why something cannot be done.
- They have the ability to join a meeting and know who really is in charge, and has the knowledge and ideas, rather than those theoretically in control.

References

Adair, J. (1983) *Effective Leadership*. Pan Books, London

Bartol, K.B. and Martin, D.C. (1994) *Management*, 2nd edn. McGraw-Hill.

Fiedler, F.E. (1967) *A Theory of Leadership Effectiveness*. McGraw-Hill.

Ghiselli, E. (1971) *Explorations in Management Talent*. Goodyear.

Gray, J.L. and Starke, F.A. (1988) *Organizational Behavior: Concepts and Applications*, 4th edn. Merrill Publishing.

Handy, C. (1985) *Understanding Organisations*, 3rd edn. Penguin Books.

Hersey, P. and Blanchard, K. (1982) *Management of Organisational Behaviour: Utilising Human Resources*, 4th edn. Prentice-Hall.

Luthans, F. (1976) *Introduction to Management: A Contingency Approach*. McGraw-Hill.

Megginson, L.C., Mosley, D.C. and Peitri, P.H. (1989) *Management Concepts and Applications*, 3rd edn. Harper Row.

Robins, S.P. (1994) *Management*, 4th edn. Prentice-Hall.

Stoggill, R.M. (1974) *Handbook of Leadership: A Survey of Theory and Research*. Free Press.

Vroom, V.H. (1995) *Work and Motivation*, Rev edn. Jossey-Bass Classics.

Vroom V.H. and Yetton, P.H. (1973) *Leadership and Decision Making*. University of Pittsburgh Press.

Wright, P.L. (1996) *Managerial Leadership*. Routledge.

Yukl, G.A. (1981) *Leadership in Organizations*. Prentice-Hall.

CHAPTER

7

Team or group working

7.1 Introduction

Unlike most other industries, there is less opportunity for the majority of construction personnel to develop long-term working relationships with other employees. This is because by the very nature of the business, a project manager is selected for a project and whilst they may choose some of the personnel, many of the rest are provided for the project as required or when they can be freed up from another project. Add to this that most contracts will have a different client and design team members; it can seen be seen that team building and the dynamics of the group are important to ensure a successful conclusion to the project. Since construction projects are on average 12 to 24 month duration, rapid team building is essential. It would also be difficult to think of situations in the construction process where an individual could operate other than in a group scenario. The project manager relies heavily on the support and co-operation of those in the team and with the industry moving away from confrontational contracts and towards more partnership types, the team now embraces many more members. Some of these are based on site, where others are in the regional, head or design partner's offices.

Whilst the author has an affinity with the word 'team' as a concept commonly used in construction, many of the texts use the term 'group'. To start with, group will be used to discuss the manner in which people can be organised, but in the latter part the emphasis is on how the group becomes a team and what constitutes an effective team.

7.2 What is a group?

A group is two or more individuals who have a common task objective, interact with each other, are aware of the others in the group and consider they are part of the group. There are two types of groups, namely formal and informal as shown in Figure 7.1.

The formal groups are defined as:

- *Command groups.* These are the groups as shown on the company organisation chart. They have been positioned in the chart to show their relationships with others in the organisation demonstrating the chain of command.
- *Functional groups.* These are groups where personnel from different disciplines are brought together to resolve problems. Sometimes members of these groups have been trained to carry out other group member's jobs. The construction site uses functional groups to carry out many of its tasks, as the site organisation comprises personnel from different disciplines working together to construct the project.
- *Self-managed groups.* Personnel in this group carry out their normal tasks and have the authority to carry out planning, evaluation and discipline and is therefore a self-contained unit. That does not mean it is not accountable to others for its actions. A construction site team can have this character. Another example is a ship's company when at sea.

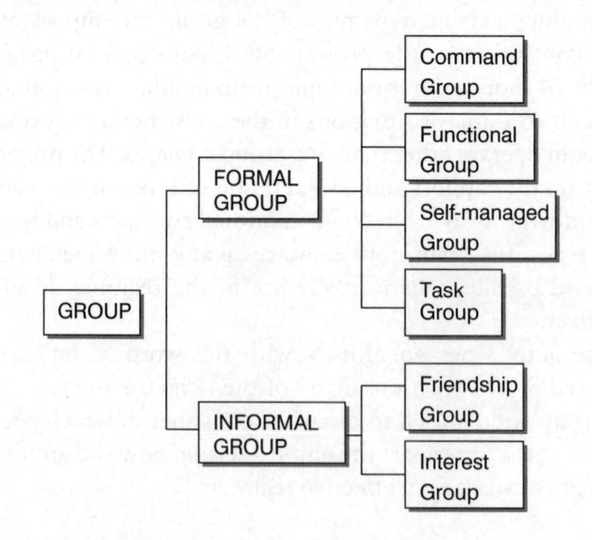

Figure 7.1 Types of work group

- *Task groups.* This group is set up specifically to solve a problem or compete a task. When completed, the group is then disbanded. Examples of this include committees, working parties and royal commissions.

Informal groups occur in a variety of ways within an organisation and are determined by the employees rather than by management. It is interesting to go into an organisation and ask employees for whom and to whom they are responsible. It is not unusual to find their answers do not tally with that drawn on the organisation chart. The effect of this is the informal group can cut across the formal structure. Understanding these informal groups can assist in using the organisation more effectively. For example, the author, when working in Hong Kong, discovered informal groups based on status were very strong, so all the chief administrators in each department regularly did, or could, meet for lunch. By speaking to each other they could readily find out what was happening in another department and report back, by-passing the formal lines of communication which were often slow and sometimes obstructive. This was an example of an interest group. Another could be the company sports club or horticultural society where personnel of different status participate.

Friendships develop between personnel, but over time promotion may separate their status within the organisation. This does not mean the friendship wanes or that they will cease communicating. One of the most important methods of informal communication is called networking, where friendships and good working relationships enable parties to cut through normal lines of communication to permit assistance. This can occur within the organisation and between different organisations. Whilst this is not an informal group as such, it does indicate the strength of informal links.

7.3 Why are groups formed?

Groups are formed either formally or informally for a variety of reasons. Formal groups are put together by organisations for good practical reasons. These include:

- The task requires more than one person to complete it. Bricklayers cannot lay bricks efficiently by themselves. They need someone to mix the mortar if they are to lay bricks all day and, generally speaking, a labourer can keep two bricklayers provided with materials, hence the gang size.
- The task requires a variety of skills, experience and knowledge. A construction site is a good example of this.

- Management hopes that by putting people together in a group it will encourage an exchange of ideas so the work process might change for the better, although much will depend on the environment the supervisor creates.
- It is hoped that group peer pressure will keep in line anyone whose behaviour is perceived as unacceptable.

The members of the group can also derive benefit, including:

- A feeling of belonging – there is a feeling of security by belonging to a group rather than being isolated as an individual. Most people find companionship positive.
- If the group is successful there is also a feeling of satisfaction and status.
- If the group works well as a group and individuals' contributions are recognised, there is a feeling of esteem.
- The group can protect a member of the group against outside pressures and threats.

7.4 Group cohesion

Generally, the greater the cohesion in the group the more effective the group will be, but there are situations where the group is very cohesive and rebels against the organisation's set objectives. Part of the leader's function is to ensure the group is both cohesive and directed towards attaining the goal. There are several factors that can affect the amount of group cohesion.

The composition of the group. The larger the size of the group the more difficult it is to control and maintain cohesion. Communication between members becomes harder and smaller factions might form which can be counter-productive. There is also the risk that some members may take a free ride. There is no magic number that determines the optimum size, as many variables come into play, but generally once the group exceeds 12 members, difficulties are likely to emerge. The compatibility of members or otherwise can play an important role. Personalities, the amount of common interest, the level of competition and conflict and compatibility of talent are issues, which determine the success of the group. Another key issue in construction is the permanence or otherwise of team members. It takes time for people to get to know each other and engender team sprit and good working relationships. The frequency of turnover of members can also affect the cohesion.

The workplace. The type of work undertaken can mean either the group is close together or separated. In the latter case communication becomes more difficult, but in spite of this if communication is good, often due its frequency, cohesion can still be high. This is important in construction, as many of the team players are not located on the construction site, such as the design team and departments like purchasing and estimating. The physical environment can also have a part to play, although much depends on the expectations and experiences of the group members. Working on a construction site is not a clean, quiet, comfortable environment, but these conditions are normal and expected. Workers in a normal office would be upset if they had to accept conditions anything like those on the site. In an office environment the debate over the pros and cons of open plan offices as against personal offices is another example of how the working environment can have a part to play. The open plan certainly enhances communication, but many prefer their private space.

Management. The type and style of management leadership will affect the relationship of the group to the needs of the organisation. If the group is successful in achieving the goals set or the goals they have set, the group will be more cohesive. Cohesiveness can also be achieved when external threats occur and the group members rely on each other for survival. Both of these can clearly be seen on outside activities courses designed to develop management and leadership skills. Delegates confronted with a strange and seemingly tough environment can be seen to bond together against, in this case the environment, as they are outside their own comfort zone, and when achieving the goal are generally highly elated.

Group development. Groups develop over a period of time as relationships mature, but not without pain. The most notable explanation of this identifies four stages referred to as forming, storming, norming and performing. Forming is when group members in a new environment, somewhat anxiously test each other out to get to know each other. As they know each other better they are more prepared to be open and express views and opinions, known as storming. Conflicts and disagreements can occur as a result, so the next stage, norming, is when members start to accept there has to be compromise and co-operation for the task to be properly executed. At this point the group can commence to perform as a cohesive group. A more detailed discourse on these is given in section 7.7.2.

7.5 Managing conflict within and between groups

It is rare that conflict will not occur within a group. Indeed it can be argued there is something wrong if there is not some conflict, as this will mean there is apathy amongst members of the group or they are so similar in nature and views, alternative opinions are not aired and discussed. There are various factors that can cause conflict. These include:

- *Task dependency.* This is when one member of the group cannot start or complete their task until another has finished theirs or when they are mutually dependent and cannot proceed unless both accomplish their respective tasks.
- *Scarcity of resources.* Limited resource availability can be a cause of conflict. These resources include equipment, support staff, office space, funding, pay increases and staff development. If two people require access to the same piece of equipment or cannot get a pay rise they think they merit, they become upset and annoyed.
- *Target conflicts.* The organisation can give different, conflicting target objectives to different parts of the organisation. For example, the purchasing department is charged with obtaining materials and sub-contractors at the lowest cost. The site then can be confronted with a selected sub-contractor who is under performing, but is still expected to bring the contact on time and within budget.
- *Personnel characteristics.* A major source of conflict occurs as a result of the individuals within the group. Loss of patience by an experienced member because a less experienced member is slow on the uptake or makes mistakes is a typical example. A member may be reluctant to own up to a mistake, whereas another believes they should. There can also be personality clashes, some members may find it difficult to work with another.
- *Communication breakdown.* It has been a common theme throughout these three books, to stress the importance of good communications. Lack of communication, or imprecise or incorrect communications, can often lead to conflicts between parties.
- *Pay and bonuses.* However much management wishes to keep remuneration to staff confidential, it will almost certainly be revealed from one source or another. There is little more disruptive when members of a group (or other groups or other organisations) are paid disproportionately to each other.

There are several methods managers can use to resolve and minimise conflict. The first stage is to anticipate likely points of conflict and take action before conflict takes place. When conflict has occurred the manager needs to evaluate the situation and establish the cause of the conflict, and having found the cause, take action to change the factors causing the problem. However, it may not be possible to change these circumstances either because it will be beyond the current availability of resources and too expensive, or it is not practical to do at that point in time. It is sometimes possible to persuade the group to look at the bigger picture rather than the objectives initially set for the group. For example, if the contract is successful and the client satisfied, there is very likely to be more work for the company and hence those employed by it, although care has to be taken in adopting this strategy, because, if work is not forthcoming in spite of a successful conclusion, then trust will be lost in the future.

Conflict is not always a bad thing as it can stimulate discussion and ideas. In certain circumstances it should be encouraged, providing it is monitored and controlled. An athlete in competition is in conflict with the competition or opposition, but the successful ones are usually in control. Lack of conflict within a group usually means that unless it is an exceptionally harmonious group, then there is a state of apathy, resulting in low levels of performance.

7.6 Team roles

This section relies heavily on the work and research Meredith Belbin and colleagues developed primarily at the Administrative Staff College, Henley and the Industrial Training Research Unit at Cambridge.

A team comprises a group of individuals all of whom have different personalities and talents. The skill of management is not just to motivate and obtain the highest collective output from them, but if the opportunity arises, put together a team that will have a range of talents and abilities that collectively will add value to the group as a whole. To do this it is necessary to establish what the talents and abilities of each member or perspective member are, and what the mix of these characteristics should be to obtain optimum performance. Individuals can be broadly classified into various types of characteristics, which if identified and understood, can be used for the betterment of the whole. It can be used to identify personal strengths and weaknesses. The latter not to be used to exploit, but to protect against by ensuring that person is not expected to perform using their weakness, but another in the group who has strengths in this area used instead. A team is only as good as the weakest link. Belbin initially defined eight different characteristics. Later this was modified to nine by adding the role of Specialist as shown in Table 7.1.

Table 7.1 The nine team roles (adapted from Belbin, 1994)

Role		*Contribution to the team*	*Allowable weaknesses*
Plant	PL	Creative, imaginative and unorthodox. Able to solve difficult problems.	Ignore details and are often ineffective communicators as they are preoccupied with the problem.
Resource investigator	RI	Extrovert, enthusiastic and communicative. They explore opportunities and are good at developing contacts.	Overoptimistic and soon lose interest once the initial enthusiasm has passed.
Chairperson	CH	Mature and confident. Make good chairpersons. They clarify goals, promote decision making and are good at delegating.	Can be seen as manipulative. They delegate their personal work.
Shaper	SH	Challenging, dynamic individuals who thrive on pressure. Have the drive and courage to overcome obstacles.	Can provoke others and hurt other's feelings.
Monitor evaluator	ME	Sober, strategic and discerning. They see all the options and judge things accurately.	Lacks drive and the ability to inspire others. They are overly critical.
Teamworker	TW	Co-operative, mild, perceptive and diplomatic. They listen, build, avert frictions and calm the waters.	Indecisive in crunch situations and can be easily influenced.
Company worker	CW	Disciplined, conservative and reliable. They turn ideas into practical actions.	Somewhat inflexible and are slow to respond to new possibilities.
Completer/ Finisher	CF	Painstaking, conscientious and anxious. They search out errors and omissions. They deliver on time.	They are inclined to worry unduly and are reluctant to delegate. Can be 'nit-pickers'.
Specialist	SP	Single-minded, self-starting and dedicated. They provide knowledge and skills, which are in short supply.	Contribute on only a narrow front. They dwell on technicalities, overlooking the 'big picture.'

Note originally the Chairperson was called the Chairman and in some texts is now referred to as the Co-ordinator and the Company Worker called the Implementer. These were changed in the interest of political correctness.

As can be seen from these descriptions in Table 7.1, there are potentially a wide range of abilities and talents to choose from if each of this type of role was represented in a team. However, it is rare for this to occur, not least of all because there is a tendency for appointers of staff to recruit and attract people similar to them. For example, the author used the Belbin analysis for several years on groups of students, all sponsored by contractors, on a construction management course. It was found that nearly all students fell into two or three of the categories described, with just a few in the others.

Putting together a team that will function well together is difficult, for example, Completers, with their intense concern for detail, are very useful in a team, but tend to annoy those who fall into such categories as Plant and Resource Investigators, hence the need for a co-ordinator to manage these conflicting personalities. A team full of Plant would equally be lost without a Completer and vice-versa. Belbin further develops the results of experiments carried out using different combinations of roles.

It is suggested when selecting a team an analysis of the types of role necessary to most successfully achieve the team's goal should be undertaken so the most appropriate team members can be selected if available. In doing this the team builder can then highlight the weaknesses and the strengths of the team as a whole so the strengths can be exploited for the good and the weaknesses realised so the objective is not let down as a result.

The titles of the team roles owe something both to historical factors and the need to avoid the preconceptions associated with the established alternatives. These could not be entirely overcome and can therefore be misread. For example, the Chairperson team role refers to the characteristics of chairpersons found in winning companies. In fact, some successful chairpersons of industrial commercial groups do not themselves adopt a typical Chairperson stance, but make their marks as Shapers, where sharp or rigorous action is the order of the day; or as Plants, where the chairperson's role is basically strategic.

Company Worker or Team Worker team-roles have tended to be undervalued because of their titles. The former has been replaced in some firms by the title Implementers, but there is much to be said for Company Worker with its flavour of someone who acts as the backbone of the company. As an alternative to Team Worker, the term Team Builder has also been used. In practice some typical Company Workers and Team Workers in the business and industrial world become chairperson of their firms. In these cases some aspects of their Chairperson behaviour are learned, although other aspects of their style are likely to reflect their primary team role.

Managing multicultural teams

As the world gets smaller and businesses work across national boundaries, and as society changes as a result of immigration, there needs to be an understanding of the effect these changes have on managing a team. The author remembers working in the Far East and being advised of the differences between the British and the locals. True, there were differences, some significant, but in reality there were many more similarities. This is important to understand otherwise there is the danger of overcompensating for each other's cultural needs. Whilst not trying to give a comprehensive and authoritative list, the examples below are meant to give an indication of the kinds of issues that could impact on running a multi-cultural team.

- *Hours of work.* Some societies believe it is a sign of incompetence to work excessive hours, others think working long hours is necessary to show enthusiasm and hope it will impress management. Others have siesta periods during the day because of the climate.
- *Mode of dress.* In the UK it is expected that people in many white-collar jobs should be dressed in suits irrespective of the temperature. In Scandinavia it is more likely to find people in similar roles dressed casually.
- *Method of address.* In some countries first names are commonly in use, whereas in others the surname is more likely to be used, especially as a sign of respect.
- *Respect and saving face.* In China age carries respect and often goes with position in the hierarchy of the organisation. The culture of 'saving face' is extremely complex. Full comprehension is difficult to achieve. For example, if you make someone lose face, then you also lose face. This has much to do with compromise in negotiations and why the signing of a contract is not always easy to achieve as their word is expected to be accepted, because if they do not meet the agreement, they lose face. Even small details such as handing one's calling card is an issue. It should be handed with both hands rather than one.
- *Authority.* Some cultures expect to be told what to do and will not question the instruction and feel uncomfortable being asked their opinion on how to carry out the task,
- *Religion.* This is significant issue for many reasons. Different faiths have different modes of dress, attitudes to women or need specific prayer times set aside. There are the different holidays such as Christmas, Ramadan, Passover, Diwali, Chinese New Year and so on, and there are issues associated with fasting and days of worship.

7.7 Management of change

Whilst there are good reasons for understanding the processes of managing change, in many instances if it is necessary to manage change, then it is too late and the opportunity has probably been missed. An organisation should be set up and managed in such a way that change can naturally be accommodated by the culture of the people within the organisation.

However, there are circumstances where it is inevitable and, in particular, when personnel are promoted or have to change jobs. In these cases the individual is put through a transition stage, which if not managed correctly, can often lead to an unsatisfactory outcome or failure. This happens often and is referred to as the Peter Principle, which states that 'every employee rises to his or her level of incompetence', or put another way, 'everyone rises to at least one level above that at which they are competent'. The effect of this is either the business does not operate to its full potential because of incompetence, or there is an unacceptable turnover of staff due their frustration or stress.

7.7.1 The transition process

This process is divided into several stages many of which the reader may have already experienced, when changing schools, going to university, moving to another job, getting married or having children.

Stage 1 – Immobilisation or shock

When first going into a new work scenario there is often a sense of being overwhelmed. The previously held view does not match reality, hence the feeling of shock. Before starting the new job, the vision of the job is too positive and it is only when seeing the reality does one see the limitations and difficulties.

Typical statements indicating this might be: What am I doing here?; This isn't the job I thought it was; Did I really apply for this job? I want my mum!

Stage 2 – Denial of change

This is a temporary retreat into false incompetence where one tries to minimise the change or trivialise it. There tends to be a reversion to previously successful behaviour. If this goes on too long then problems start to arise, as the previous behaviour becomes dominant. If an employee stays at this stage then the Peter Principle comes into play as they perform badly because their behaviour is inappropriate in the new situation.

Typical statements are: In my last job we did it this way; This situation is very similar to my last job.

It is not that the previous experience is not valuable, and maybe the way it was done in the last job might not be wrong, but it is the clinging on to past experiences as a determining factor in solving new problems that is symptomatic of the reluctance to move on.

Stage 3 – Incompetence and depression

There is a lot of frustration at this stage. The individual feels as if they are floating, because of the realisation of change, and finds it difficult to know how to cope with the new situation or relationships.

This is a very important stage in the process as without realising that change is necessary, they cannot move on and develop. Current values, attitudes and behaviours need to be challenged to permit people to cope with the new situation. There is a danger that the individual bottles this up and often organisations don't offer to help or even appreciate there is a problem. After all, they have employed the person to do the job. They offer a 'sink or swim' approach, which does not help the situation. Indeed, it can stop the real learning process and hence the development of the individual and the acquisition of knowledge.

However, this 'suffering' is important as the individual learns from it. There is a school of thought that suggests that unless one has a trauma or a major set back, the chances of being successful in life are reduced. The parallel being that the jolt to the system in both cases motivates the individual to strive for recognition or success.

Typical statements are: I'm not sure what to do; I'm confused; Can someone tell me what is happening here?

Stage 4 – Accepting reality

Prior to this point everything has been about hanging on to past values, attitudes and behaviours. At this stage there is a realisation of the actual situation and the relief and pleasure generated as a result of letting go of the past. Also there is a desire and readiness to experiment with change, optimism for the future, and a new-found ability to cope.

Typical statements are: I can see where I was going wrong; I will try again and get it right; Why didn't I think of that before?

Stage 5 – Testing

At this stage the individual tries new approaches and behaviours as a means of resolving problems; looking for new ways to overcome the transition. They become much more active than before showing high levels of energy, but also anger and frustration as the testing process progresses.

Of course as a result of this new-found enthusiasm, but relatively little experience, many more mistakes are likely to be made. Many of the experiments will turn out to be blind alleys. This behaviour needs to be encouraged and the mistakes not penalised otherwise the individual will once again retreat to earlier stages. It is only by experimenting that new and effective ways can evolve.

Typical statements are: We should be doing it this way; It didn't work, let's try another way.

Stage 6 – Search for meaning, internalisation

At this stage the individual is looking to find the meaning or reason for why things are different, trying to understand the activity and causes of anger and frustration. It is only when they have experienced the anger and frustration they can attempt to make sense of it and why it has occurred. This is usually quite a reflective period. These reflections establish the real meanings of the events that have occurred and the values that have lasted or those that need to be discarded.

Typical statements are: I now understand why that worked and this didn't; I see where they are coming from now.

Stage 7 – Integration

This is the final stage. The transition is now over. Conditions appear stable as the changes are integrated into the experience of the individual. The individual will have developed newer and better ways of dealing with the job and confidence and self-esteem will have risen resulting in increased performance.

Typical statements are: I feel like a new person; Well I'm glad it's over, but I learnt a lot; I understand where I was going wrong, but have also learnt where my true strengths lie.

The above are generalisations as to the way individuals go through the change process and of course there are overlaps between the stages, with some progressing much faster than others. It is, however, important to realise that individuals have to go through these seven stages. Further, if management is aware of the implications of these stages and at the same time watches for, anticipates and takes the appropriate action, the speed of these changes and hence higher levels of performance will be realised sooner and the likelihood of staff leaving are also reduced.

7.7.2 Team development

It is suggested that there are four distinct learning stages in the development of a group of individuals to form a more cohesive group. These are known as forming, storming, norming and performing.

Stage 1 – Forming or testing

Forming stage is a testing stage when individuals in the group become acquainted with each other. They tend to feel inhibited and so are polite with each other, impersonal in conversation, guarded in disclosing information (some will be reluctant even to offer opinion) and they will be watchful of each other's behaviour and conversation.

Stage 2 – Storming or infighting

During the forming stage fundamental differences between the individuals are unlikely to emerge because of testing each other out and because the group will tend to concentrate on completing the tasks at hand. However, once the members of the team have gained confidence, then a certain amount of infighting is likely to occur.

The leadership of the group may come into question. The initial leader, appointed at the forming stage, may be less dynamic and skilful than expected. The group members may demand a change of leadership to maintain a high level of performance, which the leader may resist. The change of leadership may also be requested not because of skills, but because style, values, etc., are considered unsuitable or a more powerful member of the group may be making a bid for leadership. This in turn may split the group into different factions.

This effect of this is infighting, even if the leadership issues are resolved one way or the other, there will still remain tensions and differences of opinion resulting in feelings of animosity and bitterness, making it difficult for the new leader to function. Also, as the group is storming different members may prefer different approaches to completing the task. This may mean certain members may feel unwanted and redundant.

Personality differences also have an effect at this stage increasing the likelihood of tension. If these tensions are not resolved, the conflict and confrontation could become the normal behaviour of the group. The problem with this is that even though the personality clash may only affect two or three members of the group, the tensions created can be disruptive for all members. If these tensions are not resolved then confrontation will become the norm.

Whatever the cause of the infighting, the behavioural pattern of the group is characterised by conflicts and divisions. Individuals who feel unmotivated leave the group and some of those left feel trapped. Others may opt out of the group entirely just concentrating on the tasks set rather than the relationships within the group.

Stage 3 – Norming or doing

The speed at which the norming stage takes place, largely depends on the style and personality of the group leader as well as pressures on the group to produce results. The sooner the group tries to fulfil certain task goals the sooner it will break out of the infighting stage.

By doing tasks, norms of behaviour and professional practice are established and barriers between members of the group begin to disappear as a result of exchanging views, ideas and experiences. However, tensions will still remain so the group of part of it may fall back into stage two. This can occur several times and, in extreme and unsuccessful groups, they never progress past this stage.

Stage 4 – Performing or identity

Breaking out of stages two and three is no easy matter. Much depends on the effectiveness of the leader who needs interpersonal, counselling and listening skills so they can act as a third party (consultant) to warring group members. It is important that group members identify with the group mission or purpose. A skilful leader is able to shape a meaningful identity for the group, which is essential for a move into stage four.

Once in stage four the group becomes cohesive. They are more supportive of each other, share information and ideas and tolerate each other's differences. They use each other's strengths and talents to a greater degree, are more resourceful and flexible in their approach to problem solving and task performance. It begins to perform as a genuine team. Problems can arise and a drift back to stage three if not properly led, especially as time goes on and new members replace other members of the group.

References

Bartol, K.B. and Martin, D.C. (1994) *Management*, 2nd edn. McGraw-Hill.
Belbin, R.M. (1994) *Team Roles at Work*. Butterworth-Heinemann.
Belbin, R.M. (1996) *Management Teams*. Butterworth-Heinemann.
Dyer, W.G. (1994) *Team Building: Current Issues and New Alternatives*, 3rd edn. Addison-Wesley.
Freeman-Bell, G. and Balkwell, J. (1993) *Management in Engineering*. Prentice-Hall.
Gray, J.L., Starke, F.A. (1988) *Organizational Behavior: Concepts and Applications*, 4th edn. Merrill Publishing.
Mullins, L. (1989) *Management and Organisational Behaviour*, 2nd edn. Pitman & Sons.
Peter, L.J. and Hull, R. (1969) *The Peter Principle*. Morrow.
Robins, S.P. (1994) *Management*, 4th edn. Prentice-Hall.
Thomas, K.W. (1977) Toward multi-dimensional values in teaching: the example of conflict behaviors. *Academy of Management Review*, 2: 484–90.
Thomson, R. (1993) *Managing People*. Butterworth-Heinemann.

CHAPTER

Human resources management

8.1 Introduction

Personnel management, as it was once referred to in the 1950s, the 1960s and early 1970s, was primarily concerned with ensuring sufficient and appropriately qualified employees were recruited into the company to satisfy the needs of management. Apprenticeships took care of skill training needs and the larger companies organised a certain amount of in-house training for line and service managers. They would also have a responsibility for safety issues and sometimes were involved in developing industrial relations strategies. It was not unusual for personnel managers to be recruited from the military. Ranks such as major and group-captain were common because of their experience in administration and their availability, as they often retired from the services aged around 40 years old.

However, in the late 1960s and early 1970s there were significant changes in legislation, encouraging companies to take a much more detailed look at the ways they recruited, what work conditions were provided, and notably, how employees were dismissed. Examples of these acts of legislation included the Industrial Relations Act 1971, Sex Discrimination Act 1974, and Health and Safety at Work etc., Act 1974. Companies found themselves attending tribunals for breaches of industrial law, often losing the case because they had not procedures in place to ensure employees were treated fairly.

Whether or not this legislation coincided with a change in society's attitude to how labour should be treated can be debated. The fact remains that companies started to take notice of the type of labour they employed and how they used and looked after it. It became important to 'beef up' the personnel function, give it a voice on the board of directors, and introduce many more systems of control into the organisation. It is perhaps significant that towards the end of the 1980s and early 1990s the term human resources management (HRM) started to be used rather than personnel management. An appreciation, perhaps, of the changing attitude and the realisation that labour is a resource, and a very complex one at that. Companies aim to obtain at least 95 per cent efficiency from a piece of plant, increasingly we try to keep the wastage of material down to a minimum and these days even consider recycling. It is, therefore, reasonable to attempt to obtain the highest output from employees, be it manual or cerebral to ensure the company is competitive and sustainable.

As a demonstration of the difficulty of the task, consider drawing up a shortlist of twelve from Table 8.1, which the reader could consider the most important issues and conditions an employee might want from their employment. This is not a definitive list. The selection made could be for a manual work or for a manager. If the list is different for both, consider why this is and should they be?

After completing this, ask the question: what are the four most common grievances you have experienced in your workplace or have read or heard of in the media? Although it would be too simplistic to draw any conclusions from this brief exercise, it is worth noting that when asking students to carry out these two exercises in groups they were able to answer the latter question

Table 8.1 Possible employment requirements

Adequate wages	Responsibility	A chance to contribute ideas
Security	Safe working conditions	Satisfaction
No Saturday work	Promotion prospects	Long hours
Work away from home	Variety of work	Short hours
Long holidays	Work close to home	Regular average bonus payments
Large wages	Sports club	Recognition of one's work
Music whilst you work	Good supervision	Transport to and from work
Job satisfaction	Acceptable work mates	Quality output in one's work
Work requiring no thought	Clean working conditions	Canteen
Work requiring thought	Fair treatment	Bonus payments
No responsibility	Friendly work colleagues	

quickly, with little dispute and generally all coming to the same conclusions. Whereas in trying to answer the former, the groups had lengthy debate and their conclusions usually were at variance with other groups.

However, the function of the human resources management unit is not just about providing an appropriate working environment for employees, but also about planning for and meeting the staffing needs of the corporate plan, which can include laying off personnel as well as recruiting and retraining, ensuring the organisation complies with and is aware of the current legislation, has training plans in place, recruits the best personnel for the job, advised on pay structures and pay rates, anticipates and deals with industrial relations problems and provides a save and healthy working environment.

8.2 Strategy and organisation – human resource planning

Human resource planning is part of the corporate business planning process. It is all very well for senior management to decide to expand, contract or diversify the business, but if it is impossible to provide the appropriate people to cope, the plan will fail. If the business is to contract this will mean redundancy and there are all sorts of implications here to the reputation of the business and to industrial relations. There are also financial implications. It is therefore essential that the HRM function of the organisation is consulted and has an input into this decision-making process.

Human resource planning is influenced by exterior and internal factors, all of which need to be understood in carrying out the board's requirements in the corporate business plan and advising the board when developing the plan.

8.3 External factors affecting human resource planning

Increasingly, clients, especially local authorities, are requiring that as part of the contract a relatively high percentage of the labour employed on the contract must be recruited from within the area. Other clients require, as part of the contract, that the contractor trains staff the client is to employ. This has cost and resource implications for the HRM department. When working in another part of the world there are further complications, mainly cultural. Other interested parties may control labour, and this may involve the criminal fraternity and there is the issue of paying cash to others for services to be provided. In the UK this may be classed as corruption, but can be normal practice elsewhere. It is important that the HRM department is aware of the differences in the way staff and labour are employed in these situations. This is one of the reasons why companies working in other countries partner with a local contracting organisation. Wimpey, for example, partnered with John Lok, a local contractor in Hong Kong, when building the Hong Kong and Shanghai Bank's head office.

Legislation is constantly changing and this has a direct effect on the company which needs to be aware of current and proposed changes initiated at Westminster and from Europe. This can be concerned with hours of work, flexible working, health and safety, maternity and paternity leave, minimum wage, holiday arrangements and trade union membership. Legislation in other countries may be different also.

The demographic changes need to be understood. Whilst the total population is increasing slowly and the split between male and female is changing marginally from year to year, the age profile is altering more rapidly with an increasing older population and a decreasing younger one. This has an impact on recruitment. Add to this the fact that the government has planned for 50 per cent of the school population to be able to go to university and higher education, there is a shrinking pool of young less-qualified people from which trade and labourers can be recruited. The image and reality of the industry working outside in often inclement weather, in dirty conditions and with less than ideal welfare facilities does not encourage many to consider the industry as a career, adding to the difficulty in recruitment. It is noted the number of women entering the industry is small and this is a potential recruitment possibility.

The availability and quality of labour, both professional and trade, impacts on the business, especially if expanding or moving into other areas to work. This can be affected by the economic cycle and the amount of construction work being conducted now and in the future. Construction has been and still is being used by governments as an economic regulator as procurement of public works can be stopped as a means of controlling their expenditure and vice versa. This is particularly problematic if a boom in construction is permitted rather than a steady increase. Fortunately, in recent years the latter has been the trend. There is no doubt that there has been a reduction in the availability of properly trained tradesmen although the influx from the extended European Union has alleviated this to some extent. Whilst much of this labour will not be directly employed but sub-contracted, the HRM department, in conjunction with the procurement department, needs to understand what the labour market is likely to be. Qualified construction staffing is likely to be a cause for concern as it relies on the continued output from our universities. The number entering certain disciplines is variable, much reflecting the public's interest. For example, there was a significant increase in interest in careers in construction after Jeremy Clarkson's presentation on Brunel in the BBC's '100 Greatest Britons' series. However, a further significant issue is the age profile of the current lecturing staff which is closer to retirement age than graduate entry, meaning there is a likelihood there will be insufficient qualified staff with industrial experience to educate the construction professionals in the future.

Design trends can also have an effect. For example, as has happened before, a trend towards building steel-frame multi-storey structures results in skill shortages in this trade. The effect of this is excessive wage demands. There are continual changing methods of construction, making certain operations quicker, requiring less human resource and in some cases less skill. There is also a trend to move site production into a factory, where quality and output can be controlled in a secure weather-protected environment. How far this will develop is still to be seen. Historically, as in the industrialised building systems of the 1960s, prefabrication has had a limited success for only a short period, but the indicators are that this time it is more likely to succeed.

The wages offered have to be competitive with other industries and other companies. Whilst there is a national wage agreement settled each year between the employers and the trade unions, this does not mean every company pays that amount. They may well offer more by increased bonus earning opportunities, plus rates to attract labour working in the same area of the country, and accommodation and travel expenses. This can become more complicated if the increases and incentives attract people from one region of operation to another. Heathrow Terminal Five project was an example of this where tradesmen were offered reportedly in excess of £50,000 a year.

Currently, the trade unions are nowhere near as powerful as they were. However, it should be noted there have been considerable problems in the past in certain areas, notably Liverpool and London, both then hotbeds of militancy, much of it very politically motivated. This resulted in countless strike actions to the extent some companies declined offers of work in these areas. Whilst at the time of writing there is moderate calm, it should be remembered that a few miscalculations by government can create the right environment for unrest to be fostered and there will be always some ready to capitalise on these failures.

8.4 Internal factors affecting human resource planning

The HRM department has to take note of the impact on staffing levels when senior management restructures the organisation. Periodically organisations can be centralised or decentralised, departments merged or divided, work functions moved from one location to another, or the business expanded or contracted. There can also be takeovers and mergers. Whenever significant changes occur in the organisation, it is not unusual to find staff are in the wrong place and/or in the wrong numbers. Relocation of staff is not always easy as many have domestic criteria important to them, such as children being at a good school in their GCSE year and spouses having their own career.

The skill profile of the work force needs to coincide with the work to be carried out. Achieving this balance may mean making staff redundant, because there is an excess of that skill, or retraining to adapt to the current needs of the business. If retraining is not an option or will not fill all the requirements, then recruitment will be needed.

The business plan may have an impact on the organisational structure of the company. It may not be able to effectively cope with the changes. This means a new structure has to be developed and this can affect the relationships between staff, lines of communication and levels of command, responsibilities and authority. The impact for the HRM department will be in then resolving unrest, morale issues, retraining, new pay structures, redundancy and recruitment. Further, if the number of staff changes either up or down, or new offices are opened, there are support functions to be considered, such as health and safety. These may not be located at the new premises, but the existing support teams have to be refocused to deal with the new situation.

One of the problems associated with rapid expansion is the amount of money it is necessary to pay new staff. If there is a shortage this will almost inevitably mean they may be offered more than staff already in post. However one attempts to keep salaries a secret, the truth will out eventually. This can cause great resentment for obvious reasons. Equally in times of a glut of applicants, paying below that of staff in post will also cause friction.

Figure 8.1 takes account of the external and internal factors and demonstrates the three stages of human resource planning.

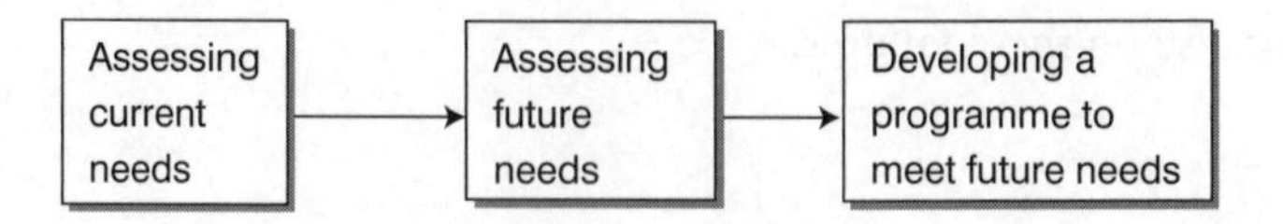

Figure 8.1 The three stages of human resource planning

8.5 Sources of potential personnel

Before investigating the selection procedures, it is worth considering where job applicants are likely to come from. Table 8.2 adapted from Robins (1994) gives a broad indication of these sources.

8.6 Recruitment and selection of employees

Recruitment and selection occurs at all levels of an organisation from the site labourer to the chief executive. Each will require a different approach, but in essence the same question has to be answered. Can they do the job? In all cases it is important to decide precisely what is required. Sometimes this is relatively simple, but it can be quite complex. It is not enough to advertise for a labourer. What skills do they need? Do they have experience in laying paving, drains or concreting? Are the joiners experienced in formwork, first fix, or second fix, as each requires a different kind of skill? Whilst it would not be necessary to write a job description in these cases, simple questioning on the type of work they have recently carried out would usually suffice, but would not be for foremen, office staff and all those in middle and senior management.

8.7 Job analysis, description and specification

Until an analysis, description and specification of the job is produced, there is nothing to compare the applicants against. It is not necessarily the same as that used for the post just vacated. Over a period of time jobs can change, either because the job itself has changed or, as is often the case, the person in post has developed the job to carry out more functions as they become more experienced. It is not unusual to discover that if the person has been in the job for a long time their knowledge and understanding is such that it may be necessary to divide the role into two or more jobs or reallocate some of the work to others. Equally it may be found that others in the team could carry out all the work and the job has, in effect, become redundant.

The first stage is to analyse the job to establish its main duties, the contacts with others as part of the normal routine, the knowledge, skills and abilities needed to carry out the role, and the working environment in which the job is conducted, including the type of plant or equipment to be used. It is also necessary to establish where in the organisation the post is, and to whom and for whom the post is responsible. This enables the HRM department to produce a job description that is a statement of the duties, specific requirements needed to carry out the job and the working conditions. From

Table 8.2 Sources of potential personnel (adapted from Robins (1994))

Source	*Advantages*	*Disadvantages*
Internal search	Low cost; improves morale of staff if they believe there to be opportunities; candidates familiar with company and vice versa.	Limited supply; may not recruit best available candidates; danger of perpetuating the status quo and not bringing new ideas from outside.
Advertisements	Can be targeted at specific groups; can be placed to cover a wide catchment area including overseas.	Often many applicants are unqualified for the post.
Employee recommendations	Current employee gives candidate knowledge about organisation; can generate strong candidates; employees know whether or not candidate will fit in.	Tends to attract people of similar type to the person recommending; danger of nepotism; may not increase the overall mix.
Private employment agencies	Wide contacts; they carry out the screening.	Can be expensive; agency may require up to six months salary as payment of fee.
Job centres	Free service.	Candidates tend to be unskilled and untrained; often lower calibre, the better candidates not using the service.
Temporary agencies	Fills temporary needs; fills peaks and can be released when workload drops.	Expensive; usually limited to a specific role.
Universities	Academic ability is measured; interviews can often be conducted at university; expected salaries are usually known beforehand.	Academic ability is not necessarily a measure of employability; expectations often different from reality and what is on offer.
Schools	Large number of candidates.	Since so many go onto university, the majority remaining have only limited qualifications and aspirations.

this a job specification, or a statement of the qualifications required, can be written stating the skills, abilities, previous work experience needed and the minimum education standards required.

The job specification is divided into two broad categories: the tasks to be carried out and the skills and aptitudes needed to accomplish them. These are divided into essential and desirable categories. Essential being those standards prospective candidates have to meet so they can function in the job and desirable being those that would be an advantage or could be taught when the candidate is in post. This document is invaluable at the interview stage as it permits all members of the panel to be working in a consistent manner when deciding to whom to offer the job.

The factors making up these would normally include the following:

- *Experience and past performance*. This is probably the best measure of the candidate's ability as they can demonstrate a track record for the suitability for the post. It would always be subject to taking up of references to confirm that what the candidate has written is correct, although this can normally be teased out during the interview phase. The level of experience needed should be clarified.
- *Technical skills*. Most posts require some level of technical skill from using a computer to operating machinery. The levels of requirement are usually easy to measure and specify, and can be tested either from past experience, certification or tests on the day of the interview.
- *Education*. The candidate producing the original certificates of their achievement demonstrates this. It is important to note that photocopies are unacceptable as these can easily be doctored, although even the originals can be modified and not spotted except by the experienced. The levels required can readily be specified. However, there is a danger of asking for higher qualifications than are needed. The risk is by appointing an over-qualified person, they will soon become disillusioned and look for another post elsewhere, or just use it as a staging post until a more suitable appointment turns up.
- *Communication skills*. This covers a wide range of abilities and skills as are discussed in Chapter 11. In writing the specification, the writer should be very clear what is required. Testing for it is complex. Certain aspects are covered in past experience and technical skills, but the more personable skills have to be tested at interview (section 8.8.7).
- *Personality traits*. These are important for two reasons. First, will the candidate be able to work with the others in the group, and second, how will people outside of the organisation view them? Measuring them will be a subjective issue, but the interviewers must have a clear understanding of what they are looking for.

- *Health, energy and stamina.* Is the candidate physically and mentally up to the job? At one extreme, footballers are given strenuous medicals before a transfer is agreed, whereas in many jobs the company might consider no form of medical check is required. Many companies do ask for a statement from the candidate of their health, not least of all to see if they have any special needs which will have to be catered for. This is important because of disability legislation. Further, as stress becomes more understood and accepted as a health issue, the candidate's ability to cope with the pressures of the job needs to be ascertained.
- *Interests.* Whilst not always necessary or essential, depending upon the post, knowledge of the candidate's interests is often an indicator of their overall personality, attitude and aptitude. They can give a picture of how rounded the person is and whether or not they are likely to want to be relocated in the future, which could be important.

It is rare for a candidate to have all the attributes required for the post so as the specification is being developed the question must be asked which are essential for the job and which are desirable. Linked to this and a determining factor as to which are essential or desirable is to decide which requirements can be learnt on the job and which will require special off the job training. Training costs money and the decision will be informed by how much it will cost, whether or not the organisation has the resources internally to carry it out and how long it might take. The latter being important depending on how quickly the person in the post needs to fully cope.

8.8 The selection procedure

8.8.1 Advertising

It is sometimes said the ideal advertisement should be compiled in such a way it attracts but one candidate who is a perfect match for the job. A slight exaggeration, as it is important to be able to select at interview from several candidates so comparisons can be made. However, a too loosely worded advertisement can attract many applications from people who are totally unsuitable to fill the post resulting in wasted time by those sifting through the applications, not to mention the time spent by those seeking employment. The basis of the advertisement is constructed from the job specification discussed previously.

Where the advertisement is placed has to be carefully considered as it needs to be read by potential candidates, and it is also expensive. It will depend upon the type of post: a secretary will be recruited locally, whereas the managing director might be found nationally or globally. Construction professionals are likely to read the journals and newsletters of their institutes/institutions or the construction press, Those in education will tend to look in *The Times* Education Supplements or *The Guardian*'s weekly education section. However these are not exclusive and potential personnel may be looking in a variety of different places. For example, who are the readers of *The Big Issue*?These days there is increasing use of the Internet. Due to this, if the advertisement is to be placed in more than one newspaper or journal it is important to have a reference code on each one, requesting that the prospective candidate quotes it in their application. This is important, as a later review of the applications will indicate where they saw the advertisement. This indicates whether or not money spent in placing adverts was effective.

The advertisement should include instructions on how the applicant should apply. This could be to send a letter of application and curriculum vitae, or to write in for details and an application form. Some companies ask candidates to complete the application form provided on the Internet.

8.8.2 Head hunting and agencies

An alternative decision is to go to a recruitment agency. The main reason for doing this is to obtain temporary personnel. It is a well-established practice in many industries, such as warehousing, where the labour force required fluctuates considerably during the year. For example pre-Christmas and the January sales often require a greater throughput of goods. Sub-contractors can avail themselves of this service, but increasingly main contractors are using agencies for the temporary recruitment of site engineers. When recruiting senior managers, it is sometimes of benefit to use agencies to head hunt for this expertise. They are in a specialist area and know the appropriate people that could be recruited this way. They will approach prospective personnel directly and then put them forward to the client for interview. This is an expensive method as the agency fee can be quite high. The advantage is they have been personally selected and are more likely to fit the job profile than results through advertising.

8.8.3 Application forms

The purpose of an application form is to have standardised information provided by the candidates in the same sequence. This enables those selecting candidates for interview to make comparisons fairly and reduces the chances of litigation by candidates who believe they have been discriminated against.

It also improves consistency if more then one person is involved in this stage of the selection process.

It is advisable to have an application form for all posts. However, the composition of the forms can vary depending upon the post being advertised, but in truth for most of the junior, middle management and administrative roles in construction, the same form can be used. There are standard application forms for graduate vacancies. Depending upon the role, a letter of application may also be appropriate. The application form can either be hard copy or electronic, with the candidate completing it on the Internet and sending it back to the employer.

The information to be solicited in the application form will include personal details about the person such as age, address and contact details; dates of education and qualifications, both academic and professional; industrial experience in chronological order, or the reverse, to include dates, roles and brief experience details; a general section to permit the candidate to indicate outside work interests and any other details the candidate would wish to bring to the employer's attention; and finally names, positions and roles of referees. There is still reluctance on behalf of many referees to act upon electronic requests for references. This is because of concern as to legitimacy of the request in handing out personal information about third parties.

8.8.4 Selecting for interview

The application forms have arrived and the next stage is to select from these the candidates to be shortlisted for interview. If the response is inadequate or the applicants unsatisfactory, steps must be taken to re-advertise. This may require revisiting the composition and content of the advertisement, the places it is posted and even the salary being offered.

It is important to realise personal prejudice can affect this selection if steps are not taken to overcome it. This can be largely overcome if two people select independently of each other using a third to assist if, when they come together, they cannot agree. There are various ways of going about this depending upon the numbers of applications received. A typical method is for each of the two independent selectors to mark and separate the forms into R for reject, P for probable and Q for maybe. Within the latter two the selector should then place them in order of priority. Their decision should be based on comparison of the completed application forms against the job specification. The two then get together and make the final selection for interview.

8.8.5 Preparing the interview

It is important to remember the interview process is not a one-way process, where the employer is selecting a candidate to fill the post. The applicant, unless desperate for employment, is also selecting the company. Whilst the prime objective is to select the best person for the job this can only be achieved if the candidates are put at ease as much as is possible so that their true selves are exposed. Information about the post and the company culture should be provided to allow the candidate to make an informed decision as to whether or not they wish to accept the job. There is little point in going through this process only to find out after the candidate has been working for a short time, the job is not as they expected.

The first stage of this process is to get the candidate to the interview and to deal with other housekeeping issues, such as how far the candidates have to travel. If the candidate has a great distance to travel it may be appropriate to interview the candidate later rather than first thing in the morning, or offer to provide overnight accommodation and subsistence, the former booked for them if necessary. Clear instructions should be sent on how to get to the company offices and where within the organisation they are to report. Remember the candidate may be coming by rail, bus, air or private car so clear directions and parking information should be addressed. Candidates should be made aware of what expenses they are entitled to and ideally provision made to either collect them on the day or have an expenses form sent with the interview details so that it can be rapidly processed on completion of the interview. The amount involved can be a significant amount of money especially if they have travel long distances and may be an issue if they are currently out of work.

The receptionist and security should be made aware the interviews are taking place so the candidates can be taken care of on arrival such as offering light refreshments if appropriate, indicating where the toilets are and taking and storing their coats They need to be provided with a list of those expected and the times they are to be interviewed. A space should also be allocated for waiting candidates.

If members of staff are required as part of the process such as escorting them around the premises, giving talks about the organisation and its work and ethos, they should be well briefed in advance as to their responsibilities.

8.8.6 Preparation of the room

The layout of the room is important to create the right atmosphere for the interview. Providing an informal approach will make the candidate feel ill at ease if they would normally expect to be sitting at a conference table facing the interviewers on the other side. This could result in some of the objectives of the interview being lost.

A decision needs to be made where the other interviewers are to sit relative to the controlling interviewer, and the use of name plates can assist in communication between the them and the interviewee. The candidates should not be positioned so they have direct sunlight in their eyes and the room should be clean, tidy and fit for purpose. The seat provided should be functional. One too comfortable will make the candidate feel uneasy and an uncomfortable one distracts them.

It is essential instructions be given out that there must be no interruptions during the interview. Providing cups of tea or coffee are also ill advised as the candidate finds difficulty in finding times to drink and will almost invariably not imbibe at all. It is an unnecessary distraction. Refreshments, if provided, should be in the collecting room for applicants.

If as part of the interview the candidates are expected to make a presentation, then the appropriate equipment should be provided and technical support staff available in case anything goes wrong.

8.8.7 Types of interview

Interviewing is a skill that has to be learnt. Many organisation insist that any staff member asked to take part in the interviewing process should have been on a staff development course prior to being put in a real situation. There are a number of different ways the interviewing and selection process can take place. These are the interview panel, selection boards and group testing. Some form of testing might supplement all of these.

The interview panel is the most common in the majority of organisations. The composition of the panel will vary depending on the nature of the employment, the level and complexity of the post to be filled. They tend to be small in number. For middle management it would typically comprise someone from personnel, the head of section the post is in and perhaps their immediate supervisor. If it were a technical post (accountant, planner, quantity surveyor) it would be normal for someone at a senior level in that discipline to be in attendance as well.

Selection boards are much more formally structured and found in public institutions such as the civil service, universities and the health service. The composition may be laid down in the organisation's procedures. They tend to be larger than the interview panels, which does bring with it the disadvantage of a committee making the decision rather than those closely involved with the role to be filled. The members will usually also include personnel from outside of the immediate department to ensure a fair decision is made. So, for example, a board set up for a university lecturing post will include professors from another department. The danger is they will ask questions of insufficient depth if not controlled by the chair of the board. The author remembers being asked by a professor of computing about his domestic plumbing, somewhat off target from the debate that had previously taken place about the differences between the education needs of civil and building undergraduates. It was suggested he look in the yellow pages for a plumber.

Both of these types of interview are in depth and should be designed to encourage a two-way flow of information between the candidate and the interviewers. In both situations the interviewers should have been briefed on what is required and the areas each should carry out their questioning so that there is a balanced approach. Whilst it may be appropriate to have a series of common questions for each applicant, there should be flexibility to allow the interview to go in any direction the chair feels is appropriate. However, this should not detract from the reason for the interview in assessing whether candidates meet the criteria of the job specification.

Group selection is both expensive and takes considerable time, not just in the duration of the event, but also in the preparation. It is a method used by some large organisations and especially the armed services in officer recruitment and selection. The basis is to bring six to eight candidates together in a group (there could be more than one group in the case of the armed services) and put them through various exercises, scenarios and tests working together as a team. These will be supplemented by personal interviews and candidates may be asked to make presentations. It is a mechanism to observe the levels of leadership, communication, group working and interaction skills, initiative, logical thinking, confidence, listening ability, inventiveness and ability to accept criticism, all often under pressure. Observers who rate and rank them against the criteria just mentioned, continually assess the candidates.

Testing can take varying forms and their use is in part a function of the job requirements. These forms include:

- Physical testing is used when the fitness and health of a person is paramount to their contribution to the organisation. A professional sportsperson would fall into this category where the testing would be vigorous and thorough. However in today's workplace, the physical and mental health of the candidate is important to stand up to the rigours of the workplace and medicals are sometimes required as part of the process. This might appear to discriminate against disability, but legislation covers this and protects against this, providing the candidate has declared their disability at the time of application.

- Ability tests are used to test mechanical skills such as working a machine or driving a vehicle. It would be normal to test a typist for their ability to use computers and various software packages, both for speed and accuracy. They can be used to test mathematical, linguistic and spatial skills, as well as sensory skills such as listening and vision.
- Personality testing is increasingly used by larger organisations to measure personal characteristics such as their ability to work with others, whether they are very detailed in their work or more cavalier in wanting to complete the job as quickly as possible. These tests usually ask the candidate to select from a series of two statements and assess which they consider they are most like. For example, 'I am hard worker', or 'I am a fast worker'. From the responses the assessor produces a personality profile. It should be noted there are varying views as to the validity of these types of test and they should always be read in conjunction with the findings from the formal in-depth interview. Depending upon the test, the person making the assessment may have to be qualified and be a member of an appropriate professional body such as the Institute of Psychology. It should be remembered there is an ethical dimension in assessing the personality of an individual.

Finally, there is the one-to-one interview usually used in employing tradesmen and general operatives on the site. The person carrying this out should be qualified because there are many pitfalls in not complying with current legislation especially if the applicant believes they have been rejected on the grounds of gender, age, race, trade union membership or disability.

8.8.8 Conducting the interview

The chairperson should agree with all the interviewers how the interview is to be conducted, the sequence of questioners, the areas to be examined, the approximate duration of the interview and that all have the correct paperwork with them, this includes the candidates' application forms and the job specification. They should be reminded they should be testing the candidates' abilities against the categories listed in the job specification and they should look beyond these and be prepared to make a judgement about the personality of the candidate. These include their appearance, humour, confidence, accuracy of response and so on.

When the candidate sits down, all the members of the panel should be introduced by name and position in the company. It is usual to open the interview with a few pleasant remarks and ice-breaking questions such as 'was your trip here all right?' and minor personal questions so as to relax the candidate and a reminder of the purpose of the interview. It also means the candidate gets used to the style, pace and approach of the interview. They should be told when the decision to appoint is to be made. Sometimes this is on the day, in other cases it will be within a few days.

Other issues the chairperson should address when controlling the interview are the direction and standard of the questions so the purpose of the interview is maintained and its pace is within the timescale allotted. It should be remembered the candidate will probably be nervous and allowances made for that fact. The questions posed should be clear, concise and direct and open-ended. They should give the opportunity for a considered response and allow the candidate to open up and elaborate. Generally the more they speak the better, providing the length of the response is controlled by the chairperson if it is in danger of going on too long. If there is any doubt about the clarity or accuracy of a response this should be addressed. The panel should assess whether or not the candidate wants the job. This can be done based on the enthusiasm shown or by simply asking the question 'if you were offered this post would you accept it?' The candidate should be given the opportunity to ask any questions they have. The interview should be firmly but politely brought to a conclusion, thanking the candidate for their attendance and, if appropriate, resolving the issue of their expenses. Above all, all members of the panel should remember the most important and most difficult thing is to listen.

There are actions members of the panel should not do when interviewing. They should not use the interview as a platform upon which to air their views and knowledge, nor should they dominate the conversation, lose their temper or argue. Debating an issue, though, is acceptable. They should not allow their personal like or dislike of the candidate cloud their impartiality and hence influence their judgement, nor should they jump to conclusions as to the appropriateness or otherwise of the candidate for filling the post until the analysis is done at the end of the interview. The candidate should not be interrupted during their response unless they are clearly missing the point of the question or are dragging it out. The questions asked should not be loaded so as to solicit the answer required. If it becomes clear the candidate is totally inappropriate for the job, then even though a given amount of time has been allocated for the interview, it should be brought to a conclusion earlier. Promises that cannot be honoured should not be made. The interview should be brought to a conclusion and not be allowed to fade out or be inconclusive.

It is perhaps appropriate to say a few words about personal prejudices. Often when the word is mentioned people immediately think of racial and gender prejudices, but everybody is probably prejudiced against some thing or another. It can be about the way people dress, their perfume or cologne, their hairstyle, their politics, their accents, where they live, the kind of car they drive, their spouse or partner, facial hair, if they are vegetarian, where they go on holiday, their weight, if they smoke and so on; the list is endless. You cannot change a prejudice overnight and sometimes never. Often they have developed as a result of a cultural background, or lack of knowledge and understanding. What is important about any prejudice is being aware of it. Knowing one's prejudices means one is less likely to be biased against the candidate, although there is a danger of over-compensating.

The decision to appoint should be carried out as soon as possible after the interviews have taken place whilst the candidates are still fresh in the minds of the interviewers. Some chairpersons ask for comments after each candidate has been seen, whilst others wait until all have been through the process. Each candidate should be considered in turn and compared against the job specification to see if they satisfy the criteria and a list drawn up of the best candidates. From those who could fill the role, further discussions should take place about their overall personality and performance before candidates are chosen as the first, second and third choice, providing there are enough satisfactory candidates. If the decision is close, then the board may decide to reconvene the following day to make a decision after having slept on it.

Once the candidate has been chosen, issues such as the salary to be offered should be agreed. All those who clearly are not suitable should be advised immediately. It is extremely frustrating for candidates hoping for employment, to be told several weeks later they have been unsuccessful. It also does the company image no good. The first choice should be notified as soon as possible and given a reasonable amount of time to make the decision, even though they may have indicated at interview they would accept the job if offered. They have a right to make a considered judgement as it may mean moving and disrupting the family who need to be involved in the candidate's decision-making process. It is no good pressuring the candidate to make a hasty decision only to find a few months later they are not happy and resign. If they accept the reserve candidates can be notified, and if turning it down, the second choice can be offered the job.

8.9 Training and staff development

No organisation can afford to stand still, there will always be change as new opportunities and challenges arise. To cope with this staff will continually need to be trained or retrained to meet these needs The amount of training required will depend upon the role of the individual, but it is a reasonable assumption that staff will have training throughout their career. However, it must be done for some purpose that will benefit the company and the employee. The Construction Industry Training Board (CITB) was formed in the 1960s as one of 29 industrial training boards (ITBs) set up as a result of the 1964 Industrial Training Act, but by the 1980s this had reduced to only two, one for the engineering industry and the other the CITB, the latter as a result of strong lobbying by the major construction companies.

The reasons for training are to introduce new members of staff to the company, to make visitors to construction sites aware of the dangers and their responsibilities, to meet technical advances, to bring employees up to speed with new ideas and legislation, to give the employee greater job satisfaction, and to retrain for new work or on promotion.

8.9.1 Induction courses

When a new employee starts they arrive somewhat bewildered, perhaps a little frightened of what's ahead of them and maybe questioning their own confidence. The purpose of the induction course is to resolve these matters. The composition of the course will vary depending upon circumstances. In all cases, the new employee needs to be introduced to those with whom they will be working, especially their immediate supervisor and subordinates. This is not just useful for the employee, but also existing staff that need to know who this new face is and what their role is. They need to know where they are working, where the basic facilities are, such as toilets and canteens, where the stationery office and stores are. Larger organisations may wish run a more formal course to discuss the overall ethos and objectives of the company, introduce them to safety, quality and environmental policies, what to do in the event of fire and to meet senior staff, spending a day or more in the process.

On construction sites is it not unusual to give all visitors an induction course related to safety issues before they are allowed on site to ensure they are wearing the correct protective clothing and know the evacuation procedures and where not to go. Some of these courses are specially developed for specific risks such as working in tunnels and sewers and how to use and understand the safety equipment they have to carry, such as methane detectors. More information on safety and training is found in *Operations Management for Construction,* Chapter 4.

8.9.2 Training and development needs

There appears to be a link between the ability to hold on to staff and the attention given to staff development, especially from senior management.

On analysis this is not surprising, since senior management is more sensitive to the direction of the company and hence the future needs for staff to fulfil this ambition. The training needs of the company are determined by the overall corporate plan for the future. It is usually beneficial to the company if the human resource needs can be met within the company rather than recruitment from outside, although new blood from time to time bringing in new ideas can help to invigorate the business. The human resource department has to analyse the company's needs based on the corporate plan and assess the talent available that can be developed within the organisation and the gaps in their ability to cope with the work to be carried out. Alongside this, they must consider the changing work environment as new technologies and legislative requirements come into play as well as staff expectations and aspirations, which can be established through staff annual appraisals.

The next stage is to design the training and implement it in practice. The training can be carried out in a variety of ways, the most common being on the job. Here the employee learns or is coached whilst carrying out their role, usually with the help of a mentor who could be the supervisor or an experienced colleague. The mentor will assess at set intervals the progress of the employee and this will be used as basis for discussion in the annual staff appraisal. Job rotation though different parts of the organisation may be another method. Often trainee graduates are expected to spend a certain amount of time in various departments such as planning, estimating, purchasing and quantity surveying as part of their training package. This is not just to learn what happens in different functions of the organisation and how they relate to each other, but also to give the employee the opportunity to discover which types of work they best relate to and might wish to make, or not make, a career in. Another way is to have a planned progression of promotion within the organisation. Job shadowing is sometimes used where the employee stays with the senior manager for a period of time seeing what they do and how they carry out their jobs. Employees are sometimes given temporary assignments to give the company the opportunity to see how they are able to cope, as well as probably enhancing the employee's job satisfaction and motivation.

Simulation techniques are also available where the employee is subjected to simulated exercises, where they can do no harm, but their performance can be closely monitored and feedback given. The most sophisticated building simulation model has been developed at the BMSC Leeuwarden, in the Netherlands and is now licensed in the UK to be sited at Coventry. Each trainee is placed in a 'site cabin', connected to the outside world by phone and computer to the central control and the instructors. In each cabin they have the complete set of information for constructing the project such as drawings, specification, the bills of quantities, the construction programme, names of suppliers, sub-contractors and members of the design team. They are then set a series of tasks to accomplish within a set period of time. Outside the huts and only visible when exiting the cabins, is a large screen showing the site at any point in time during its construction that is relevant to the exercise being set. The trainee can use a joystick to go to any part of the site and building. There are also simulated background site noises. When in the cabin, the trainee can access information from the documentation or by contacting suppliers, sub-contractors and members of the design team who are played by the instructors. However, this process is two-way and the instructors can contact the trainees and put problems in their way. To make matters worse, various members of the site team with their problems interrupt the trainees at intervals. Actors play these roles. At the end of the sessions the trainees are debriefed. This is an excellent method of developing managers' skills and to establish whether or not a prospective manager has the ability to cope under such pressure.

The company can run courses to update employees about changes in legislation, amendments to the JCT form of contract, the impact of revised or new company policy and systems and so on. It can also put on courses to develop IT skills and to learn to operate new machines. Sending personnel on courses organised by others, or bringing experts into the company premises, as well as being provided in-house, can teach some of these. Sending executives on university courses, outside activities courses and conferences can provide other off-the-job training and development programmes which focus on management skills.

It is important in all these training scenarios that the quality and success of the training is evaluated against the criteria developed when deciding the training needs. Obtaining feedback from both the trainee and course controller/instructor, testing levels of performance, assessing any behavioural changes, can do this. The latter by observations made by the staff members' immediate supervisor.

Retirement needs to be prepared for and many companies run retirement training courses for their employees. It is very important because the change from full-time employment to retirement can be traumatic, resulting in premature death. The main issues are finance, health and what do with one's time. On the surface it appears great to suddenly have no worries and responsibilities, but the truth can be very different. On the financial level, there will be usually a significant reduction in earnings even if receiving a company pension and with many of these no longer being offered, the situation could become much worse. Many schemes provide a lump sum and this needs to be invested and used wisely. Assistance in knowing what to do is invaluable. Over the years, many people in full employment, especially if they have had a senior post, have not kept fit and this can result in health problems on retirement. Regimes of the kinds of exercise and diet need to be explored to combat the change of lifestyle. Finally the executive used to delegating work and organising others moves into an environment where there is nobody to organise except a partner or spouse. The latter may have been for years looking after the home, supporting the working member of the family and then suddenly there is significant change in both lives. Equally if both have been working similar problems occur.

8.9.3 Construction Industry Training Board (CITB)

The construction industry has changed in the last 40 years from the one that directly employed most of the main skills, such as bricklayers, joiners, scaffolders, steel fixers, crane and machine drivers, and the labouring skills such as drain layers and concreters. The mechanical and electrical services, plasterers and painters were usually sub-contracted. The result was that construction companies used the apprenticeship schemes then in place as well as a lot of on-the-job training, especially for the labouring roles, providing most of the industry's training needs in-house. Recently, there has been a steady move towards employing very few site operatives and using sub-contract labour instead. The result has been a significant deterioration in the number of apprentices trained directly by the construction companies, except in isolated pockets in the country. The CITB has become the main supply of trained operatives. It is funded by an annual levy paid by all liable construction employers. It then redistributes these monies to companies of all sizes who carry out the training in the form of a CITB Grant. In 2004, the CITB distributed nearly £100 million in training grants, which equated to £1.79 for every pound raised by the levy. This is because the CITB also distributes money obtained from other sources such as the Learning and Skills Council and The European Social Fund.

In the early days of the CITB the emphasis was on training a skilled labour force including plant operators, but in recent years there has been a movement towards management training. The CITB annually produces a publication itemising the grants available, their purpose and value. Construction businesses can then apply for them, but it should be noted these are generally paid after the training has been completed unless the training spans more than one year.

8.10 Welfare, health and safety

Employers have both a statutory and moral responsibility to look after the welfare, safety and health of all those they employ and to third parties who either visit the premises or interact in some way with the work process such as pedestrians walking past the entrance to the site, neighbours and other road users. This section is primarily concerned with welfare, as discourse on health and safety can be read in *Operations Management for Construction*, Chapter 4.

The minimum statutory requirements can be found in the Construction (Health, Safety and Welfare) Regulations (1996) and the National Working Rule Agreement. They are primarily concerned with the physical provision of welfare serviced and are summarised in Table 8.3.

Whilst these are minimum requirements, there is nothing to stop more being provided than is required, especially in the support offered to employees both on site and in the office, other than restrictions on the employer in terms of the cost of providing them. Indeed there are good arguments to suggest that by providing high standards of welfare, motivation and loyalty are increased, so it may be more beneficial to the well-being of the company as a whole.

There are further calculations to be made about the size and quantity of the facilities to be provided which is a function of the number of personnel employed on the construction site.

To the above in Table 8.3 could be added the provision of subsidised canteens. It would be expected the wages of the staff preparing and cooking the food and the food itself be covered in the selling price, but the subsidy could be in the accommodation, fuel, heating, lighting and general overheads. Consideration should be given to the provision and identification of healthy food, although there tends to be a tradition on site to eat considerable helpings of the traditional English breakfasts, chips and so on as a means of employees obtaining a high calorie intake to balance the energy used in the type of work and cold conditions experienced in the winter months. A further provision is first aid, but if someone is injured and needs to go to hospital for treatment, part of the welfare provision is consideration on how they get there and back, if they need to be accompanied and their family notified.

Car parking is always problematic in terms of space and is often a cause of unrest. Whilst in the office environment, the amount of car parking space is already determined, there should be a transparent policy in place as to the means of allocation and proper provision made for those with disabilities. On site the same should apply, but if there is sufficient space available, steps

Table 8.3 Welfare provision

Sanitary conveniences	Rooms shall be adequately ventilated, lit, and kept clean and orderly. Separate rooms shall be provided for men and women unless the door to the room can be secured from the inside. Ensure there is always an adequate provision of toilet paper. Female toilets should have an effective means of disposal of sanitary waste.
Washing facilities	Shall be provided in the immediate vicinity of every sanitary convenience and any changing rooms and be large enough for the washing of face, hands and forearms. There shall be a supply of clean, hot and cold water preferably running water, and soap, cleansing agents and towels (paper or cloth) or driers, should be provided. Rooms shall be adequately ventilated, lit, and kept clean and orderly. If showers are required for very dirty work or when dealing with hazardous materials, showers should be provided apart from the normal washing facilities. Separate rooms shall be provided for men and women unless the door to the room can be secured from the inside.
Drinking water	Drinking water shall be clearly and conspicuously marked with an appropriate sign, cups provided or the water supplied in a jet from which people can drink easily, such as a drinking fountain.
Accommodation for clothing	The accommodation should include or allow for facilities for drying clothing.
Facilities for changing clothing	There shall be separate changing rooms for men and women.
Facilities for rest	These shall include rest facilities in the rest rooms and suitable rest facilities for pregnant or nursing mothers if required and shall include suitable arrangements for food to be prepared and eaten as well as the means for boiling water.

Adapted from Schedule 6 of Regulation 22 of the Construction (Health, Safety & welfare) Regulations 1966 and Provision of Welfare Facilities for Construction Fixed Sites – Construction Information Sheet no 18 (rev1)

should be taken to ensure cars using it are not damaged. Some sites require a substantial car parking facility for the completed project and consideration should be made to programme this work in earlier rather than later if possible so site personnel can use it.

Personal support for employees is important. This can be on various levels, and some individuals are more likely to provide it because of their style of management, for it to be universal, the ethos of the company must be partial towards it and this can only happen if determined at board level. Whilst not suggesting a 'nanny state' approach, the fact is employees can have problems at work, at home and in health and do not always know how to deal with them. Supervisors should try to develop relationships with their group so employees feel able to speak in confidence to either their immediate supervisor or someone else in the organisation, so supervisors will need training on how to do this. If employees are on long-term sick or have recently suffered a bereavement of a close relative, the organisation should be aware of this and take an active interest in the employee's situation. The support given can take many forms, including paid leave, visits from other staff and advice about benefits and other support available in the local area if appropriate. These days, the provision of crèche facilities may act as an added attraction to prospective employees who might find the working hours problematic.

Social activities such as sporting events, theatre evenings, visits and trips can be encouraged to engender team spirit and involve family members. With ever-increasing pressure at work, people commuting longer distances to and from work and other interests there is a limit to their success. Some organisations have had their own sports facilities for staff to use, but these have to a large extent disappeared. They are expensive to maintain, the land was a useful asset, and there has been a decline in interest by employees.

8.11 Salaries, wages and pensions

The difference normally used to distinguish between wages and salaries is that wages are paid weekly and salaries monthly and traditionally wages have been paid to manual workers and salaries to others. The method of payment does affect how each group balances their family budget. Most people have monthly outgoings such as mortgages, council tax, insurance, utility bills and loans. So for salary earners, February is a good month for a fixed income and outgoings; for a wage earner it is a bad month as they only receive four wages instead of five as in some other months.

Setting salary and wage levels is a careful balance between what the company can afford and being able to retain and attract staff. What the company is able to pay is a function of the income, expenditure other than wages, investment in research, development and new equipment needs, and producing profit at a level that satisfies the shareholders, if a limited company.

American texts refer to this subject as compensation. You get compensated for your work and effort by receiving wages and benefits. Wages are remuneration for the time the employee has worked, such as the basic pay, overtime and bonuses, whereas benefits are such things as health and life insurance, sick pay, holidays, flexible working, pensions, retirement age allowance and stocks and share schemes. The total package of wages and benefits should be considered as compensation for work.

The management perspective

Contrary to popular belief, generally management would like to pay good wages, but in return they want a high level of output: 'A fair day's work for a fair day's pay'. This is achieved by offering incentives to increase productivity, improving management performance to motivate and increase work rates, and investment in state of the art technologies and equipment.

There can be restrictions on what can be paid. Governments will usually determine the level of pay and annual increases for public employees based on inflation rates and can apply pressure to the private sector to keep to similar limits when negotiating annual increases. They also have to pay the minimum statutory wage. Employers also have to take account of supply and demand of certain skills, sometimes paying well over the norm. They can only pay from what the company earns and if profit levels are low this will impact on salaries and wages.

Other factors affecting pay levels result from the collective bargaining at national level between the trades unions and the construction employers that determines the minimum wages and conditions for the construction employees. However, it should be noted, that since bonuses are also paid on top of this, the employer does have some flexibility to increase pay. Those employed in office and managerial roles are not covered by any such agreements as unlike in the public sector there are no laid-down pay scales. Pay therefore is more likely to be determined by comparisons with competitors within the industry.

The employee perspective

Equity is important, as individuals will compare their income and personal effort against their colleagues in the same section, others carrying out similar work in other organisations, and colleagues in the same organisation apparently at the same level, but carrying out different functions as with a planner, engineer and quantity surveyor on site. Unfortunately, supply and demand can make this difficult to achieve if, for example, there is a shortage of quantity surveyors and they are paid more than the other apparently equivalent roles. Interestingly, employees are less concerned about the pay other professions get compared to themselves, but heaven help the employer who pays more to someone of the same rank if it is perceived they do not work as hard as the others. Differentials with an organisation are a sensitive issue and can be eroded over time if not continually inspected. Much has been written and said about the pay differential between men and women and there is plenty of evidence that there still is discrimination for comparable work.

The perspective on pensions has changed in recent years as a result of scandals associated with company pensions, schemes being closed and a bar on new entrants as employers take stock of the financial implication as well as more public awareness of the volatility of the stock market. (A typical pension scheme may involve the employee putting in 6 per cent of their income and this being matched by 12 per cent or more by the employer.) Those left intact are primarily in the public sector and it is suggested that the implications of not having a guaranteed pension, and needing to make alternative arrangements for retirement, is having an effect on the way employees assess pay offers. Since working patterns have changed and many more women are at work, flexibility and holiday issues become an issue, especially if there are small children to be looked after.

8.11.1 Pay structures

There are two distinct ways in which pay structures can be designed. First, and most predominate in the public sector is fixed published scales, usually set and negotiated nationally, where the employee receives annual increments until reaching the top of the scale. There is usually some overlap between the top of the lower scale and bottom of the next higher scale as demonstrated in Table 8.4. In this case, each annual increment increase is calculated as 4 per cent higher than the one below. This maintains the differential between the grades as they progress up the scale.

This gives the employer a certain amount of flexibility when recruiting staff within a scale as they can offer the job at different points depending on supply and demand. This can cause a problem if a person appointed earlier has been offered a position lower down the scale. The main problem is there is a limit to the ability of the employer to pay a person on merit to either encourage or hold onto a member of staff if they are at the top of the scale unless there is a promotion opportunity. This can be overcome in part by having a system that permits acceleration up the scale so an employee can rise by more than one increment. Alternatively, there can be merit increments (shown in italics at the top of scales A and B) that the employer can use in special cases to reward for extra responsibility or outstanding work. In spite of these difficulties, employees take some comfort from the

Table 8.4 Example incremental salary scales

Increment point	*Scale A (£)*	*Scale B (£)*	*Scale C (£)*
1	10,000		
2	10,400		
3	10,816		
4	11,248		
5	11,699	11,699	
6	12.167	12.167	
7	*12.653*	12.653	
8	*13,159*	13,159	
9		13,686	13,686
10		14,233	14,233
11		*14,802*	14,802
12		*15,395*	15,395
13			16,010
14			16,650
15			17,317

fact that annual salary increases are negotiated nationally by the unions who have a perceived strength in the negotiation process, because of the number of people they represent and their experience, with any suggestion of locally negotiated pay deals are treated with caution.

The second method of designing pay structures most commonly found in the private sector is based on job evaluation. This approach is increasingly being used in the public sector as well. Job evaluation is the process of establishing the relative merit of each job within an organisation with the aim of creating the correct differentials between them. There are various methods available, the most common referred to as the points rating method. It selects criteria common to all jobs and rates them on a scale allocating points accordingly. Typical criteria could be experience, supervision, number and importance of decisions, responsibility, effort required, skill required, working conditions, education and training, supervision and complexity of work.

Once the criteria have been selected, a scale of points has to be devised for each. For example experience could be divided as follows:

Level 1	0–6 months	10 points
Level 2	6 months–1 year	20 points
Level 3	1–2 years	40 points
Level 4	More than 5 years	60 points

In this case 60 points is the highest to be allocated, but another of the criteria, such as a post with a high level of responsibility, may be considered by management to merit more points. How exactly the points are decided is as a result of discussion between senior management and the expert putting the scheme together. Once the scale has been devised, then each job within the organisation is compared against it and the points calculated. This is usually supplemented by interviewing staff, since their current role may not reflect what was in the original job specification as jobs often evolve. This then acts as the basis to determine the par rates throughout the organisation. Before using the system, a few experimental trial runs of some of the posts are carried out to see if the outcome appears to make sense and the pay being suggested appears rational.

This exercise should be carried out at periodic intervals, because circumstances and needs alter as work methods change, products and services are developed and the organisation structure is amended, giving and taking away responsibilities and decision-making responsibility. The problem arises when differences are found resulting in some employees now being overpaid. The question is should they be able to keep their current salary with annual cost of living increases, keep it without cost of living increases, or have it reduced to the new scheme?

The other aspect is to consider whether there is equity between what has resulted from the exercise with what is being paid in other comparable organisations. This process is called a pay survey or wage and salary survey to benchmark a cross-section of the company's jobs (usually 25 to 30 per cent) against those in other organisations. This is normally carried out using mailed or telephone questionnaires with companies who are prepared to share this information. It might appear strange that competitors are prepared to share this apparently sensitive information, but in practice most of the information is available as a result of personnel changing companies, advertisements and other sources. It is in everybody's interest to be open with this information. The information gleaned from this process is then compared with the outcome from the points rating method so the organisation can develop its wage structure.

8.11.2 Incentive schemes

It is too simplistic, and incorrect, to believe by just offering financial rewards performance will improve. As discussed in Chapter 1, many other forces generate motivation. However, there is no doubt that financial incentives can succeed especially if supported by good management practices.

Incentive schemes for site operatives fall into two categories: piece work, where operatives are paid a rate for the work done with no basic wage; and those which are in addition to the nationally agreed hourly wage rate. The former has in the past been applied to work that is simple to measure, such as brickwork, where the bricklayers would be paid so much per thousand bricks laid.

There are some basic rules that should be applied to any bonus scheme for operatives. These are:

1. Bonus should be paid in direct proportion to their performance where it can be measured, normally with no limit to earnings. That proportion could increase once performance has reached a stated level to act as a further incentive, or deceased to discourage any further increase if it is felt quality or safety standards might be compromised.
2. The method of calculation should be fair and transparent and readily understood by all. This is not just good practice, but goes a long way to eliminate suspicion by the operatives about the integrity of the calculation. In the past some schemes have been highly controversial and resulted in industrial action. For example in the post-war years, dock workers could be paid by several employers during the week depending on the ship they were working on, would have different hourly rates depending on the cargo and different bonus rates. It was difficult to check the calculation and even if it was done and there was a dispute, the ship would have probably left port so it was impossible to recheck.
3. The nature of the work to be done and the times allowed should be clearly specified. The author remembers a case when the time given for cutting a hole through a stud partition was given without a clear specification of what the work entailed. At the end of the week when the work was measured the joiners wanted twice as much because there was a hole on both sides.
4. The times allowed for work should be given before it commences otherwise there could be suspicion the times were decided afterwards and the earnings manipulated.
5. They should not be altered at a later stage unless both parties agree. Situations do arise when the times given are too tight for the operatives

to make reasonable bonus and need to be relaxed. Equally there are occurrences when the times given allow the operatives to make excessive money. This is more difficult as the operatives are naturally more reluctant to take a cut. One mechanism for doing this is to trade these off against those that need to be relaxed.

6. Allowances should be made for situations when the operatives cannot work because of reasons beyond their control such as being held up by others, waiting for materials and inclement weather. The latter is covered in part by the National Working Rule Agreement. How much these allowances should be is a mute point, as operatives would argue they should be paid at the same rate as they would have earned if they had been able to work. The employers are more likely to argue it should be a fixed rate or none at all.

The calculation can be produced in several ways. The most comprehensive and therefore most expensive to administer is the unit rate when each separate operation is given a target time. For example, collect tools and materials, fix 19mm x 100mm skirting using plugs and screws and cart away excess material @10metre run per hour; or it can be a fixed amount of time for concreting a complete floor of a multi-storey block. An alternative approach is to offer a fixed bonus if the work a person or gang is charged with carrying out is completed by a stated time. In both cases the number of hours the job is worth is based on work-study measurement from which the standard minutes are calculated (*Operations Management for Construction*, Chapter 3). The difference between the hours earned and the hours worked forms the basis of the bonus calculation, referred to as the *hours saved* method. Another approach is to use the labour element of the rates from the bills of quantities as the basis for the bonus scheme.

A decision has to be made on how much bonus should be paid to operatives and at what point in their performance should they start earning bonus. These are referred to as geared schemes. Figure 8.2 demonstrates examples of a 50 per cent and 100 per cent scheme. In the case of the former, when the operative commences earning bonus it is paid at half the hourly rate and in the latter at the full hourly rate.

This calculation is based on the assumption that when operatives are working at a standard level of performance of 100 rating (see Chapter 3), they should earn the equivalent of a third of their hourly wage in bonus. The vertical scale represents the rating performance. In this example it is assumed the hourly wage is £8 per hour. It can be seen that the 100 per cent scheme only starts to pay out bonus when the operative has achieved a 75 rating, whereas those on the 50 per cent scheme receive bonus at a

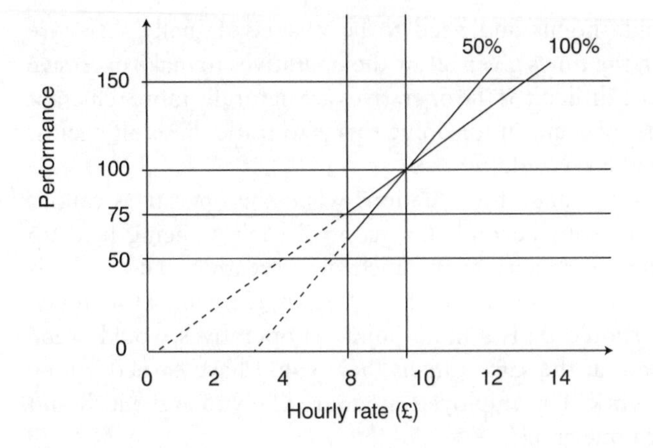

Figure 8.2 Geared incentive schemes

lower rate. However, once achieving a 100 rating performance, those on the 100 per cent scheme earn more than those on the 50 per cent scheme, management benefiting as a result, but this acts as compensation against when the operative is performing poorly for whatever reason. To permit this to happen the standard times used have to be adjusted accordingly. In the case of the 100 per cent, the standard times are increased by a third and in the case of the 50 per cent, by two thirds.

The problem with the 100 per cent is that unless the time sheets are properly supervised and filled in accurately, and the standard times and the targets are set accurately, the system becomes unworkable as either the employer or the operatives could lose out financially. Due to this, the 50 per cent scheme has generally been the most popular in the construction industry.

It is difficult to produce a bonus scheme based on productivity for operatives such as crane drivers who are servicing many operations. One of the ways of overcoming this is for them to be paid an average of all the various gangs of operatives they are servicing. Those employed on maintaining plant and equipment can be rewarded this way, but another approach is to base their bonus earnings on the percentage utilisation of the plant. In order words, their bonus increases with fewer breakdowns and down-time. However, the problem with this is that once they have achieved 100 per cent efficiency and the plant is available for work at all times, they cannot earn any extra bonus.

When decided it is not practical to have a system as above, because of the costs and experience required to run it, lack of data to produce it, or because it is felt to be inappropriate, then plus rates can be paid. This is in addition to the hourly rate irrespective of output. It is used to attract labour when there is a shortage, when the work is very complex, it is difficult to measure and devise a sensible scheme, or to overcome potential industrial problems. The latter demonstrated in Liverpool in the 1960s when because of the militancy of the trade unions, bonus disputes were closing construction sites down as a result of strike action and some companies resorted to paying plus rates in lieu of bonus.

Profit sharing has been given a high profile in the news media over the last few years because of the amounts paid to some senior executives often referred to as 'fat cats'. To put it in perspective many people are involved in profit sharing and do not receive anything like the amounts quoted. It should also be put in the context of the levels or responsibility a person has and the number of people they employ. Businesses with multi-million/billion pound turnovers employ large numbers of personnel, not to mention those employed in the supply chain. There are only a few capable of successfully managing such empires or capable of turning round one that is floundering. Unfortunately since the payouts are usually based on the success of the previous year's trading, by the time they receive it, the business profits may have taken a dip, temporary or otherwise, and the executives bonus tends to be compared with that fact rather than that of the successful year or years (some are based on more than one year's trading). The basic premise in profit sharing is that employees, not necessarily senior management, receive a prearranged percentage of the profit the company makes. A word of caution: those further down the hierarchy have no control of the bigger picture, nor know how the company is doing until the amount is given to them. If over the past few years employees always received a similar amount of money as a result of the company's success and suddenly they don't, then they can become demoralised, especially if they have come to expect it.

At least one major construction company has been working on the concept that senior management's annual bonus should be related to the safety performance of their region on the grounds that if their income is affected by the performance of those below them in this area, they will be even more determined to have an excellent safety record in their business.

Finally, there is the incentive paid as commission. This is normally associated with personnel who sell and will more likely be found in the speculative housing market. However, there are other potential opportunities for commission to be paid to those whose job it is to obtain work, but since in construction this is normally senior management, they are more likely to be signed up to a profit-sharing scheme.

8.12 Dealing with staff reductions

There are various options when having to reduce the labour and staff in put to the organisation. These are:

- Redundancy is when the employee is dismissed on the grounds there is insufficient work to justify their continued employment. This can be either compulsory or voluntary. The danger of the latter is good employees see this as an opportunity to obtain a lump sum payment knowing they will be able to start work quickly elsewhere leaving behind the less able. There are procedures laid down in law, which may be enhanced by the employer, that require a given period of notice and so many weeks or months wages/salaries depending on length of service. It should be noted that since it is the job that is redundant, it cannot be refilled as soon as the employee has left and thereby used as a mechanism of getting rid of someone the company does not like.
- Early retirement is when an employee is asked to leave employment before reaching retirement age. The employee is usually offered an enhanced pension as an incentive to go. They may also be entitled to redundancy pay. They may be allowed to be re-employed on a part-time basis, usually providing their earnings do not exceed the difference between their salary prior to retirement and the amount of pension being received.
- Temporary layoff is when the employee is laid off for a period of time and then re-employed at a later stage if still available for work. There may be some incentive paid for them to remain available for work, but much will depend on the implications of the impact on state benefits being received whilst out of work.
- Natural wastage is when personnel leave and their posts are not filled. This is as a result of staff retiring, leaving to find work elsewhere, being promoted, or taking voluntary redundancy.
- Redeployment is when staff are moved elsewhere within the organisation as their post becomes redundant. They can be moved laterally to another post, which may require retraining, promoted, or in some cases, demoted. The latter may mean the employee has to take a reduction in wages, or it can be 'ring fenced' until such time as the employee moves out of that post and the new person is employed at the lower rate.
- Reduced working is when the number of hours or days worked in the week is reduced. In the former, this may be accomplished by stopping or reducing overtime. The latter is more drastic, but is more preferable both to the employer and employee than reducing the daily hours below

the normal working day. There can be legal implications in reducing the working week because of the contract of employment.

8.13 Industrial relations

Having good industrial relations in the workplace is one of the main cornerstones for having a productive work force. Management has the responsibility to create such an environment, but from time to time need reminding of their obligations. The trade unions are the main means of providing this. The term trade union often brings up the spectre of the 1950s and 1960s when there was a combination of very poor management in many industries and a very politicised section of some trade unions resulting in much of the industrial strife of this period. Much has happened since then, partially due to the employment and trade union legislation enacted in the late 1970s and 1980s by the Conservative government, but also a change of attitude by both employers and the unions as they adapted, not just to the legislation, but to the needs of society as a whole. Whilst there is still opposition to unions in some quarters, many see advantages in their existence, not least because it is a much more convenient way of negotiating terms and conditions nationally for their industry, and, at site level, dealing with one representative rather than many is more effective and efficient. A good steward will only bring legitimate grievances to management's notice. The unions on site have the right to elect stewards to represent members' interest and a safety representative.

The trade unions represent the interests of their members not just in wage negotiation, but also the welfare of the members and their families and much of their good work is not reported as it is not newsworthy. Whilst there appears to a conflict of trade union pay demands and employers looking after their own and shareholders interests, both parties realise the need for a satisfactory outcome that suits both parties.

The Union of Construction Allied Trades and Technicians (UCATT) is the principal union in the construction industry with a membership of 125,000 (2006). It was formed in 1971 bringing together the Amalgamated Society of Woodworkers, the Amalgamated Society of Painters and Decorators, the Association of Building Technicians and the Amalgamated Union of Building Trade Workers. It comprises 10 regions each of which meets monthly to discuss regional matters.

The Transport and General Workers Union (T&G) has been involved in organising building workers for a very long time, but chose not to merge with UCATT, no doubt due to its wide involvement in other industries.

Similarly the GMB has members working in the construction industry as well as in other industries.

The contractors are represented at the negotiating table by the Contractors Confederation, which was formed in 1997. Its eight constituent organisations that made up the confederation were: Major Contractors Group; National Contractors Federation; National Federation of Builders; Federation of Building Specialist Contractors; Civil Engineering Contractors Association; House Builders Federation; British Woodworking Federation and Scottish Building. The new Construction Industry Joint Council Working Rule Agreement, replaced the Building and Civil Binding Agreements in January 1998. Besides Acts of Parliament, this governs conditions of employment within the construction industry. Amendments to it are agreed at intervals, currently every three years, between the Construction Confederation and the main trade unions representing construction workers, namely UCATT, T&G and the GMB. It is also the forum for other discussions and negotiations.

The Construction Industry Joint Council comprises 22 seats, 11 belong to the confederation, 6 to UCATT, 3 to T&G, and 2 to the GMB. UCATT acts as the lead negotiator for the trade unions. Normally at the negotiation, there are only five members from the trade unions and five from the Confederation. There are three tiers of negotiation except in the case of emergencies (a potential serious strike for instance). These are at local, regional and national levels. This means theoretically that an employee or employer can bring a suggestion or a grievance to the local panel and have it decided there. If agreement cannot be reached the matter can be transferred to regional level and then upwards to national were the decision is binding.

Tactics are a crucial part of the wage negotiation process. Both sides of the negotiation assess their own and the other side's position. The contractors are looking at their ability to pay an increase, taking account of the current and projected market and their profit requirements and that of their shareholders. They look at the trend in wage demands on all employers, in and out of the industry, in the public and private sector, as well as the rates at which they have been settling. Government signals are also noted. From this they can build up a picture of what the range of settlement might be with the unions. The unions are doing exactly the same, taking note of pressures from members in the union and what their demands and expectations are. From this both sides calculate the best settlement hoped for (BS) and what they believe to be a realistic settlement (RS). The contractors decide the maximum payout they are prepared to make and the unions decide the lowest offer they are prepared to concede. These positions are called the fallback positions (FBP).

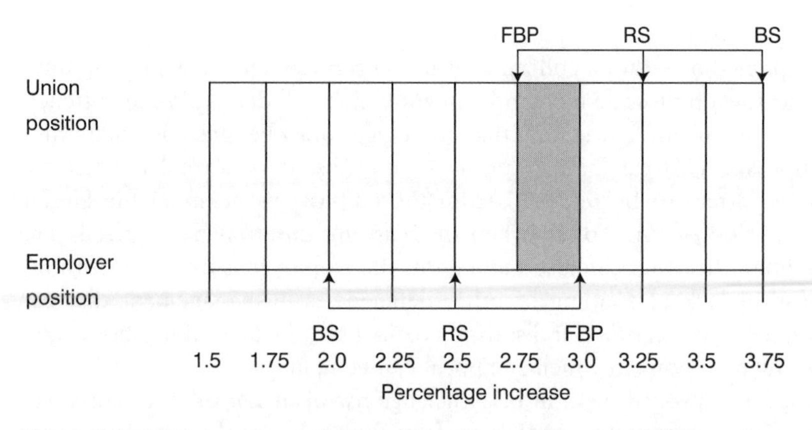

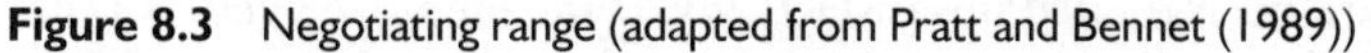

Figure 8.3 Negotiating range (adapted from Pratt and Bennet (1989))

Figure 8.3 illustrates the position the two sides might take. The employers will lay down their best settlement terms which the unions reject before putting forward their ideal settlement terms. As can be seen in the figure these are far apart and obviously unacceptable to both sides. There will be some discussion giving the reasoning behind their positions and the sides will withdraw to 'consult their members' agreeing to meet on another occasion. It may appear on the surface to be a waste of time, but to the skilled negotiating team it indicates the kind of deal the other party might settle for. They meet again and gradually the two sides move towards each other until hopefully a deal is struck, usually somewhere in the grey shaded area.

8.14 Disciplinary and grievance procedures

Organisations should have published disciplinary and grievance procedures. These outline the procedures that the employer and employee should follow in the event of a problem. In the former, much has been determined by the various pieces of employment legislation and case studies in industrial tribunals. A typical procedure would first define what was unacceptable behaviour, such as poor performance or gross misconduct, the latter usually meaning fraud, theft, deliberate damage to property, violent behaviour to others, drunkenness, accessing pornography on the web, or flagrant disregard of standard procedures and regulations. Employees can be dismissed instantly for gross misconduct. Unacceptable performance can range from under-performance, such as poor quality work, bad timekeeping, use of the web for personal reasons, and excessive sick leave, although it is necessary to tread carefully in this instance. In these cases there will be a series of warnings issued. These normally start with a verbal warning, then a written warning, and then possibly a disciplinary hearing at which the employee may be accompanied by either a colleague or union representative. Warnings issued will include a time scale for improvement which will depend on the offence. There is much case law stating that employees must be given adequate time to improve. This can be as long as 12 months. It is normal practice for warning letters to be removed from the employee's personal file after a stated period of time, often when the warning duration has expired. The disciplinary hearings can also highlight if the employee is under performing because of lack of training or other deficiencies in management that can subsequently be rectified. If dismissal is called for, the procedure should give a specific time limit to which an appeal can be made.

Grievance procedures can be either against dismissal or if an employee believes they are been treated unsatisfactorily. If the employee believes they have unfairly dismissed, before going to an industrial tribunal, they can follow the appeal route as outlined above. If they feel they are being unfairly treated, they can follow the standard grievance procedure. A typical procedure is first to discuss the matter with an immediate supervisor. If no satisfactory outcome occurs, the issue can be put to the departmental manager, and if no response is given, a written complaint can be made to the human resource management department when a meeting is set up with senior management for the matter to be resolved. The employee would expect to have union representation at this stage, although some organisations may permit this at the meeting with the head of department. If the company cannot resolve the matter, the union could ballot their members on strike action after which it would go into the regional and then national stages of dispute resolution.

8.15 Factors influencing industrial relations

There are many factors that can affect industrial relations. Some are generated within the company and others are outside its control. Figure 8.4 demonstrates the external and internal influences on industrial relations.

If management does not organise, plan the sequence of work to give continuity, and have the correct amount and quality of materials as required for good productivity, employees not only cannot earn reasonable bonuses but become demoralised as they cannot carry out their work sensibly. It is frustrating to be taken off work before completion or have to hang about waiting for materials and instructions. Management's attitude in the way they treat employees, their disregard or lack of interest on safety and quality go to create animosity or apathy from the employees. If facilities such as the toilets and rooms for eating are dirty and unhygienic it is a demonstration of lack of care and interest by management. Perceived or actual low levels of pay and

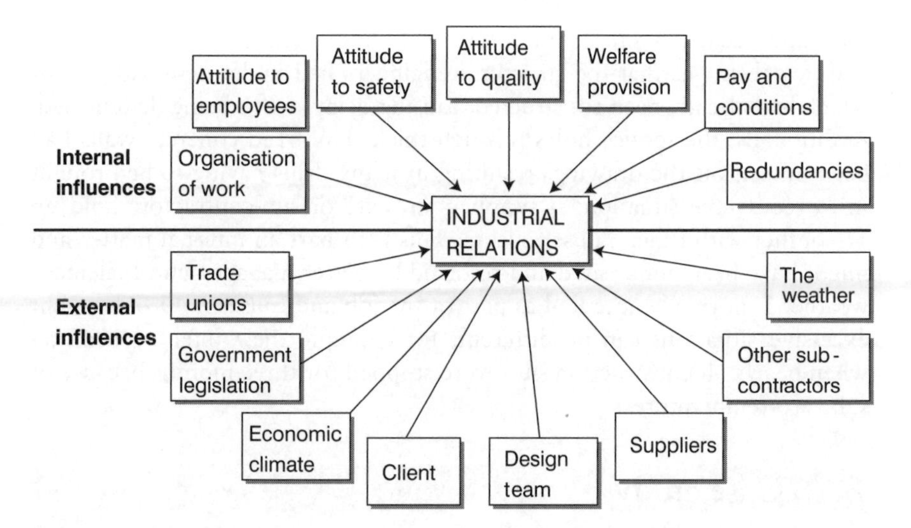

Figure 8.4 The external and internal influences on industrial relations

reward for effort have a significant impact, but also disputing the minutia on expense claims has a long-term impact on relationships. Whenever there are rumours and then enactment on redundancies, morale inevitably takes a tumble, but how it is handled can mitigate these effects.

Trade unions generally wish to resolve issues rather than have confrontation. The problem they have is with a rogue steward on site who can have a strong influence on fellow members especially over the bonus paid. This is not the problem it was when the contractor employed a high percentage of the workforce and the whole contract would be stopped as a result of industrial action. These days, strike action is more likely to be threatened as a result of breakdowns of national negotiations and the risk of this happening is low. Government employment legislation can cause industrial unrest as was shown in the 1970s and 1980s, but the more serious issues have now been addressed, so it suggested the risk here has receded. Government decisions can have a profound impact on the economic well-being of the country and hence the amount of work being carried out. Uncertainty of employment impacts on industrial relations. However, a more likely impact on the state of the industry can come from influences outside the country, such as the cost and availability of fuel, changes to the economies of other countries, war, and terrorism.

If a client changes its mind during the contract and makes significant changes to the project, this causes disruption and frustration to employees, especially if they have to demolish work they have already completed. Equally, if the design team makes errors which require remedial work, the same occurs. A site in the 1970s in London had a major dispute, the main catalyst was that the structural engineers had under-designed a floor which had already been constructed, causing delays when being demolished. Additionally, the service holes through thick reinforced concrete walls had been missed off the drawings resulting in many drills having to be brought in to rectify the situation. If suppliers are late or sub-contractors hold up or conflict with others, upset occurs. This is in part an internal matter and might have been anticipated and resolved by better management. Inclement weather is normally accepted as part of the job and causes no problem, but excessive situations can be different, for example the winter of 1962/63 when nearly all construction sites were stopped for three months because of sub-zero temperatures.

8.16 Case study

This aims to highlight some of the issues that should be considered when making management decisions to avoid disputes. Written in **bold** are some questions the reader might consider before moving on to the next section. Names have been changed.

Late Thursday afternoon, the precast concrete factory production manager received an urgent message by telephone from the construction site that two extra loads of components were required that evening. The production manager confirmed they would be delivered as requested. At this point there were eight men working in the stockyard, it is almost 4.30p.m. and no overtime had been planned for. The production manager contacted the stockyard foreman instructing him to get these deliveries loaded immediately and sent out. Ten minutes later the foreman came back and said that of the gang size of three required, two had agreed to work, but the third had refused.

What advice you would give to the production manager?

The production manager asked the foreman which man had refused to work. 'Lever. I thought that this was the chance to get rid of him once and for all. When I told him to work on, he swore at me and said that he wouldn't, so I told him if that was his attitude then there was no more work for him here and he would have to pick up his cards on Friday [a euphemism for 'you're fired' based on the fact that National Insurance stamps were physically stuck on a card which employees took with them from job to job]. Did I do the right thing?'

The production manager thought this over and said, 'Well it's not exactly the way I would have handled it, but its done and we must give him his cards on Friday. However, we've still got the problem of loading these trailers. Go and ask Chadwick to stay on. I think that he'll be agreeable as he always seems to need a bit of extra money.'

The foreman returned five minutes later to say Chadwick also had refused to work. He said also Lever had contacted the union steward about his dismissal and the steward was 'kicking up hell outside' and was waiting to see the production manager. Chadwick and Lever were also outside.

Write down your answers to the following:

- On the evidence so far, and on that alone, what advice would you give to the production manager now?
- What is the real problem here? Is there more than one problem, and who, if anyone is at fault?
- What are the irrefutable facts so far?
- What other facts would you like to have?

For the reader's information some additional facts to assist in the decision making and recommendations are:

- Lever has been with the company for nearly five years. He is an extremely versatile operative and is able to work all types of cranes in the stockyard to a very high level of proficiency.
- The stockyard foreman has never liked Lever because he finds him a difficult man to control.
- It takes three months to train a competent crane driver.
- Lever does not like working overtime and occasionally takes time off.
- Chadwick is thought to be a first-class worker, willing, 'works all hours'.
- The stockyard hours tend to be erratic because of the type of work.
- It is payday.
- Chadwick lives seven miles from the works and brings two men with him in his car each day.
- The production manager's opinion of the steward is that he is a reasonable man.

Answer the following questions now:

- Write down the facts as you now see them.
- What are the problems now?
- Who is to blame?
- What action(s) would you advise now?

The original problem was and still is, the fact loads have been promised to the site and are now not going to be delivered as promised. The production manager made a hasty decision without fully thinking out the consequences and should have told the site he would ring them back, checked with the stockyard to see if they could comply and then advised the site he could not comply with their request. To be fair, this type of request was not unusual, and the decision to send loads under such circumstances had been made many times before. The result of this poor decision was the site was angry because they had personnel waiting for a loads that did not arrive. Lever was dismissed unfairly, Chadwick, who had also refused to work but who had not been dismissed, was treated differently, there was an angry union steward outside, a foreman who had used the occasion incorrectly to dismiss someone he did not like, and a production manager whose fault it was in the first place. The author is well aware of the situation that had been created, because he was the production manager!

The solution was that the site was informed and fortunately was very understanding. It was agreed Lever should work elsewhere in the factory. It was discovered in the course of the conversation he always had wanted to work in that part of the plant. The foreman was taken to one side by the factory manager and rebuked, as was the production manager, who in turn learnt a major lesson in management.

8.17 How to deal with a problem

It would seem appropriate at this juncture to write a few words on problem solving. The most important thing is to have a clear understanding of what the reason is for solving the problem and what is expected as a result, i.e. the objective.

Stage one is to obtain the facts. Review the record; discover any customs and rules that might apply, as these may restrict solutions or have to be changed in some way; and then talk with all the parties concerned to get feelings and opinions to obtain the full story. Remember people can be reluctant to tell the full story, either because it might cause trouble for another, or they wish to tell what they think the interviewer wishes to hear.

Stage two is to analyse the facts and information. From this several actions might be possible, but before taking a decision, look at any restrictions that could affect choice. These could be company policy and the requirements of good management practice such as safety, quality and the environment. The options available could also create further problems and have an effect on the labour force, in terms of redundancies, moving personnel elsewhere and retraining, as well as the production process, as looking at one scenario may have an effect on another part of the production process. Finally, is the action decided upon going to achieve the original objective? It is important in this thinking process, that whilst possible outcomes may come to mind, it is wrong to jump to conclusions until all the facts and possibilities have been examined. There is a parallel here to the examination process outlined in the work study (*Operations Management for Construction,* Chapter 3).

The third stage is to take action. Is the originator of the decision going to take the action, is help required in enacting it, or is it to be delegated to others, and if so, what information do they require to carry it out? Does anybody else need to be informed of the decision, such as the immediate supervisor? The timing of the enactment can also be important. Making people redundant the week before Christmas or trying to change a production process at the peak period may not be appropriate. If the action to be taken is unpleasant, the buck should not be passed to others to do, or others blamed for the decision.

Stage four is to periodically monitor the success or failure of the action taken. A decision needs to be made on how soon after action has been taken the effect will be monitored, and the frequency thereafter. This could include changes in people's attitudes and relationships. Above all, has the action taken achieved the initial objective and, if not why not, and can any modifications be made to make it work.

8.18 Causes of unrest

As a final thought, included is a list of causes of unrest advanced by employers and employees back in 1914 abstracted from the Commission on Industrial Relations, *First Annual Report, 1914*, pp 19–22. Whilst some of the points are now dated, interestingly many of these attitudes still persist.

Causes of unrest advanced by employers

1. normal and healthy desire for better living conditions;
2. misunderstanding and prejudice, lack of conception that interests of labour and capital are identical;
3. agitation by politicians and irresponsible agitators;
4. unemployment;
5. unreasonable demands arising from strength of union;
6. labour leaders who stir up trouble to keep themselves in office;
7. inefficiency of workers, resulting in ever-increasing cost of living;
8. rapidly increasing complexity of industry;
9. sudden transition of large numbers of immigrants from repression to freedom, which makes them easy prey to labour agitators;
10. universal craze to get rich quick;

11. decay of old ideas of honesty and thrift;
12. misinformation in newspapers;
13. too many organisations for combative purposes instead of for co-operation;
14. violence in labour troubles;
15. sympathetic strikes and jurisdictional disputes;
16. boycotting and picketing;
17. meddlesome and burdensome legislation;
18. the close shop, which makes for labour monopoly;
19. financial irresponsibility of unions.

Causes of unrest advanced by employees and their representatives

1. normal and healthy desire for better living conditions.
2. protest against low wages, long hours, unsanitary and dangerous conditions.
3. demand for industrial democracy, and revolt against the suppression of organisation.
4. unemployment, and the insecurity, which the wage earner feels at all times.
5. there is one law for the rich, another for the poor.
6. immigration and fear of over supply of labour.
7. existence of s double standard, which sanctions only a poor living in return for the hardest manual labour, and at the same time luxury for those performing no useful service.
8. disregard of grievances of individual employees and lack of machinery for redressing same.
9. control by Big Business over both industry and the state.
10. fear of being driven to poverty by sickness, accident or loss of employment.
11. inefficiency of workers on account of lack of proper training.
12. unfair competition from exploited labour.
13. the rapid pace of modern industry resulting in accidents and premature old age.
14. lack of attention to sickness and accidents and the delays in securing compensation.
15. arbitrary discharge of employees.
16. blacklisting of individual employees.
17. exploitation of women.
18. ignorance of social economics on the part of employees.

References

Bartol, K.B. and Martin, D.C. (1994) *Management*, 2nd edn. McGraw-Hill.
Geary, R. (1962) *Work Study Applied to Building*. The Builder.
Health and Safety Executive (2004) Provision of Welfare Facilities at Fixed Construction Sites. Construction Information Sheet No. 18 (rev1). HSE.
HM Government (1996) The Construction (Health, Safety and Welfare) Regulations 1996.
Langford, D., Hancock, M/R/, Fellows, R. and Gale, A.W. (1995) *Human Resources Management in Construction*. Longman Scientific and Technical.
Megginson, L.C., Mosley, D.C. and Peitri, P.H. (1989) *Management Concepts and Applications*, 3rd edn. Harper Row.
Pratt, K.J. and Bennet, S.G. (1989) *Elements of Personnel Management*, 2nd edn. Van Nostrand Reinhold.
Robins, S.P. (1994) *Management*, 4th edn. Prentice-Hall.

CHAPTER

Managing stress

9.1 Introduction

During the 1950s and 1960s it was not unusual to hear people talk about occupational illness and accidents as acceptable as an occupational hazard. Times have changed and it is now totally unacceptable to put workers into work environments that are dangerous or, at least, to provide appropriate protection. However, until recently stress has been considered by many to be macho and part of the job, and yet now we are all aware that it has the potential to cause serious illness and even death.

Increasingly there has been a change in attitude and management is becoming aware that stress can be self inflicted by individual on themselves for variety of reasons, or can be imposed by management often without them being aware they are doing it. Besides any ethical reasons for monitoring and controlling levels of stress there are good pragmatic reasons for doing this as the amount of work days lost, the reduction in performance, reduced morale, the implications of fatality and staff looking for employment elsewhere, caused by stress is significant on the successful running and profitability of the company. The HSE (1995) estimated that 6.5 million working days were lost due to stress-related illnesses costing employers £376 million and society £3.75 billion. By 2004 this had increased to 13 million working days, employers' costs to £700 million and society £7 billion. If a person is ill and off work for any length of time, the hole left has to be filled and this can be difficult and disruptive. Clearly if this illness is stress related and could have been avoided, it makes sense to take precautions to stop this happening in the first place.

Further there are legal reasons also which notably include:

- Health and Safety at Work etc., Act 1974 where employers have a duty to ensure, so far as it is reasonably practical, the health of their employees at work;
- Management of Health and Safety at Work Regulations 1999 which include provisions for stress assessment, prevention and training;
- Working Time Regulations 1998 and the Working Time (Amendment) Regulations 2002;
- Protection from Harassment Act 1997.

Stress should not be confused with pressure, the effect of which might be stress. Many people enjoy reasonable pressure as it can be stimulating and motivating, but it is when this reaches the stage when the individual feels they can no longer cope with the demands placed upon them it becomes work-related stress. The effect can be either physical, such as heart disease or minor illnesses, to the psychological such as anxiety and depression.

9.2 Is there a problem?

The first stage in stress management is to establish facts. These can be found by using both quantitative and qualitative methods. Quantitative measurements can be achieved by checking the amount of sickness that has been taken as a percentage of the whole organisation, narrowing it down to specific areas of work and individuals or small groups. In the latter case this information has to be used sensitively, carefully and analytically, as it could be work-related, other health problems or just taking time off for personal reasons. The turnover of employees is another potential indicator of stress in the workplace, but could be due to the overall morale of the company and levels of pay. In any case this knowledge would be useful to management. Finally, inspecting trends in productivity where it is measurable, may also be an indicator of stress, but should be, as with turnover, equated with other potential problems such as the quality of management and morale.

Qualitative measurements are more a function of the working relationships between supervisors and subordinates. There are formal and informal ways of obtain these facts on stress. The informal approach of daily listening to, talking to, and observing individuals' performances can be the first indicator there is a problem. However, if the relationship is strained or very formal this will probably be ineffective. Interviewing employees on return from sick leave can also highlight potential problems providing they are prepared to confide. The annual staff appraisal interview is another opportunity to discover how the employee is reacting to the work loads set and the interviewer should be trained to pick up signals of impending problems.

9.3 How the risk of stress-related problems are evaluated and who is most at risk?

Not everybody stands up to or thrives on pressure. This may be genetic or one's environmental conditioning, so there is no norm against which to make comparisons. Further, other factors outside the workplace may have an effect on the vulnerability of the employee. For example, they are going through a divorce, moving home, a member of their family is ill or has died, they have been diagnosed with a serious health problem or they a have financial problems. There are several factors that impact on the levels of stress that individuals are subjected to. They include:

Organisational culture

Positive cultures take work-related stress seriously and have good communication between employees and management, respond to any concerns, allow staff to contribute to decisions which affect them, give them responsibility and job satisfaction. All of these reflect good management practice. Added to these are recognising there is a problem and encouraging staff to talk about it, offering support, discouraging long working hours and dissuading staff from taking work home. Finally, creating a culture that permits staff to express their concerns confidentially about their colleagues with being seen as 'telling tales'.

Excessive demands

There is always the risk of overloading employees with too much work than can be accomplished within a normal working period or expecting them to carry out work that is beyond their current capability or knowledge, because it is too difficult, or they have not been trained. Sometimes employees are also asked to carry out tasks that are impossible. In these cases the conscientious employee is likely to work excessive hours, curtail lunch breaks or take work home, worry about it and suffer stress as a consequence. The exception to this is when there is a mini-crisis within the workplace, and 'all hands to the pump' are required. Stress can also occur when employees are under-worked and bored.

The working environment

It is not possible to be prescriptive here in terms of what are acceptable ranges of noise, vibration, temperature, humidity, ventilation, lighting and cleanliness in the workplace as these can vary depending on the type of work. For example, a construction site environment is different to a hospital, library or classroom. However, within each of these work environments there are acceptable and unacceptable limits which need to be considered, because if they are not it affects not just the stress people are working under, but as a result, productivity. Before the new airport was built in Hong Kong a large number of the schools were situated directly under the flight path to Kai Tak airport. Every time a plane came into land, approximately at three-minute intervals, the children on hearing the approaching aircraft would cover their ears until after the noise peak had been reached. Clearly the learning process was severely affected. In some workplaces the fear of violence is also of concern such as with the staff in accident and emergency departments in hospitals, which necessitates steps to be taken, not just to protect staff and prevent occurrences from happening, but also to give staff peace of mind and reduce their already high levels of stress caused by the very nature of their work.

Control of work

To be told what to do without being asked for an opinion on how the task might be accomplished, can be frustrating and unfulfilling and result in a deterioration of mental health. Equally, to give employees full control of their work without providing some form of lifeline in terms of either monitoring or advice can have detrimental effects. There would also appear to be possible problems associated with control of the work environment, such as the inability to open windows, but this may also be due to ineffective ventilation systems rather than not been able to have control.

Relationships with managers and colleagues

The quality of the relationships with others is fundamental to satisfaction at work. Without this, the workplace soon becomes a stressful place to be. Problems usually manifest in one of two ways: bullying or harassment. Both forms are unacceptable in an organisation. Harassment is usually defined as 'unwanted conduct' based on sex, race, colour, religion, nationality, ethnic origin or disability, whereas bullying is defined as 'persistent unacceptable behaviour by one or more individuals in the organisation against one or more employees' (HSE, 2001). The latter includes acts or threats of physical abuse, verbal abuse, insubordination, victimisation, ridicule, humiliation, libel, slander, malicious gossip, prying in to personal matters, pestering or spying. It also includes such management practices as setting impossible deadlines, excessive supervision, deliberate and unjustified fault finding, withholding information, ignoring, refusing reasonable requests for leave or training and preventing career development. Sometimes the perpetrator is unaware of the effect of their actions especially with forms of harassment, such as physical contact between employees of different gender, poking fun and patronising disabled colleagues.

Managing change

Few people like change and when it occurs usually find it stressful mainly because of the change to status quo and uncertainty of the future. Management need to clearly communicate and identify the reasons for the changes to allow employees to question and comment, and involve them in the decision-making process where it affects them. This should happen as quickly as possible to stem any rumours, which will rapidly spread, causing even more uncertainly and stress. In any reorganisation, job roles may change, and the employee may be working with different and sometimes unknown personnel, all of which should be taken into account.

Role definition

It is unsettling for employees not to have clearly defined roles or lines of responsibility above and below. Giving written terms can help, with the proviso there can be some flexibility in the interpretation and also discussing with them to ensure they fully comprehend their role, their priorities and what is expected of them. This is an ongoing process so if their role changes for any reason, they are still clear. It is important there are no ambiguities in the interpretation of the role and there is no conflict of responsibilities. For example, a site quantity surveyor may be responsible to the site manager to provide the necessary information for the manager to act upon and for their attendance, but the regional or head office senior quantity surveyor may be responsible to ensure the quality of their work is to a high enough technical and professional standard.

Staff development and support

Staff should receive sufficient training to carry out their jobs properly and this should be monitored regularly as the job may change either as a result of modifications imposed by organisational change, or natural change, which occurs in any organisation as employees take on more responsibility as they grow in confidence and knowledge. New employees should be carefully selected so their abilities match the job closely or, if not, the differences can be filled by training. On arrival they should go on an induction course. This is very important because, if designed properly, it eases the transition from bewilderment on arriving in new employment to being assimilated into the organisation.

Employees welcome being congratulated for having done a good job and this acknowledgment by management helps in relieving work-related stress. When work is not up to the standard expected, it is more profitable to offer support and guidance so performance can be improved, although there does come a time when official warnings and eventual dismissal may have to be considered. The assumption that all employees in a given job have the same abilities and talents is dangerous. It is important to establish their strengths and weaknesses so they can be allocated jobs that play to their strengths, although some weaknesses can be overcome with further training, as they may be knowledge-based.

One of the more complex issues to address is support for employees who are going through stress resulting from their personal life. This is because they may not wish to talk about it even if the company is sympathetic to these types of problems. The company needs to develop a culture where employees are aware that help is available, management is trained to both spot difficulties and offer, or be equipped to offer, assistance or refer to another party. A word of caution though, referral can also be interpreted by the person seeking help as rejection. Awareness of an employee's personal requirements, such as being a single parent, so that work patterns can be adopted to suit their needs, reduces stress levels considerably and at the same time the company benefits from a more satisfied and committed employee.

9.4 Spotting signs of stress and anxiety

Identifying stress in personnel is very difficult to do as two people working in identical situations may find different levels of stress in the job, one perhaps getting a buzz from it, the other being unable to cope. Thus it is not possible to define the job and say this is an unsatisfactory level of stress, other than engaging in good management practices as outlined in the previous section.

However, there are certain indicators and symptoms of stress that management should beware of, all of which are based on deviations from the colleague's normal behaviour. Examples include:

- *Lack of concentration.* We all have off days when it is difficult to concentrate, whether it's from feeling ill or from lack of sleep. The problem occurs when the lack of concentration continues over a period of time.

- *Loss of motivation and commitment.* This is relative to the employee's norm as some are more motivated and committed than others. Similar to lack of concentration, but where the employee no longer comes up with ideas, enthusiasm dwindles and there is an overall lack of interest in the goings on in the business.
- *Irritability.* The employee apparently overreacts to minor annoyances where before they would take it in stride or laugh it off.
- *Quality of decision making.* Dithering and being indecisive, delaying or just making obviously incorrect decisions, compared with previous performance.
- *Increase in the numbers of errors.* Linked with the quality of decision making, but also in the accuracy of the rest of the employee's work.
- *Deterioration in organisation of work.* Individuals have to organise their short-term and some of their medium-term workload as part of their job. If the level of organisation changes this is yet another indicator of loss of concentration or motivation.
- *Inability to control personnel.* If the employee is responsible for the management of others, then excess stress levels can have an effect on their effectiveness in organising and controlling others.
- *Reduction of output.* All of the above may be contributory factors to an overall reduction of output and performance. In some types of work, especially manual, this will be easily measured in terms of the amount of work done per day. In other types, it manifests itself in the speed of responses to requests for information or decisions.
- *Withdrawal.* An extreme example would be when a very gregarious personality starts to withdraw into themselves taking little part in the general office chatter or social events and a loss of humour.
- *Alcoholism.* Sudden excessive drinking is another indication all might not be well, not to mention its effect on the employee's work output and the safety of themselves and others.
- *Increase in sickness.* Stress can cause both physical and mental problems. This can result in increases in absenteeism, behavioural changes as outlined above or noticeable changes in physical appearance, such as loss of weight, drawn features, complexion changes and difficulty in breathing.

Whilst the above should not been seen as a definitive list nor should it be used as a checklist for diagnosis, stress-related illnesses are ailments and only a qualified physician or psychologist is equipped to analyse and diagnose. These are meant only as indicators for the manager to suspect all is not well and to take further action if it is felt appropriate.

9.5 How to help

If it is noted by a colleague or is noticed by management that an employee appears to have the symptoms of stress, it is important to remain calm and objective. There are various actions the manager may consider. First, to look at the activities the employee is carrying out, prioritise them, and remove the less important. Unless the situation is very bad, it is suggested taking the important and often more difficult tasks away would further inflame the situation, as the employee would suffer a loss of self-esteem. Prioritising can be done casually without the employee necessarily being made aware of the action being taken, or preferably, in consultation with the employee to agree the revised workload.

Second, preventive action should be considered by advising staff on how to identify stress in themselves, suggesting how they might manage it and, above all, creating an organisational culture which encourages them to seek help if them are unable to cope. If any of these symptoms persist, appropriate medical advice should be sought.

When stressed, the body produces more of the so-called 'fight' chemicals preparing the body for an emergency. These include adrenaline and noradrenalin that raise blood pressure, increase the heartbeat, make one sweat more and sometimes reduce stomach activity, and cortisol which releases fat and sugar into the system. This was all right when it was necessary to fight

Table 9.1 Symptoms of stress

Physical symptoms		
Headaches	Nausea	Indigestion
Cramps or muscle spasms	Tearfulness	Muscular aches and pains
Pins and needles	Susceptibility to infections such as colds	Chest pains
Sleep disturbances	Constant tiredness	Restlessness
Dizziness	Fainting spells	High blood pressure
Craving for food	Lack of appetite	Constipation or diarrhoea
Behavioural and psychological symptoms		
Poor concentration	Memory loss	Irritability
Increase smoking if a smoker	Increase drinking of alcohol/tea/coffee	Late for work
More accident prone	Withdrawal from social contact	Reduced work performance
Indecisive	Inflexible	Fear
Depression	Apathy	Lack of motivation
Low self esteem	Fear of failure	Aggression

an opponent, run away as fast as possible or carry out some super physical act, but in a sedentary office environment these chemicals act in a different way. The result of early symptoms that might occur are shown in Table 9.1.

References

Health and Safety Executive (2001) *Tackling Work-Related Stress. A Manager's Guide to Improving and Maintaining Employee Health and Well-Being*. HSE.

Health and Safety at Work, etc. Act 1974

Management of Health and Safety at Work Regulations 1999

NHS Direct (2008) Online Health Encyclopaedia. Available: www.nhsdirect.nhs.uk/encyclopaedia/index.aspx?WT.svl=nav. Accessed 14 October 2008

Protection from Harassment Act 1997

Working Time Regulations 1998

Working Time (Amendment) Regulations 2002

CHAPTER

10 Risk analysis and management

10.1 Introduction

Risk has been alluded to and discussed elsewhere in this book and its two companion volumes, but the subject is of such importance it is identified here as a subject in its own right. Risk is associated with everything we do, as individuals, where even the air we breathe can potentially harm us because of pollutants and allergens, and in the workplace. In construction it is usually considered in terms of financial risk and the risks associated with safety. The latter is developed further in *Operations Management for Construction*, Chapter 4, but many of the principles outlined here are relevant. Risk is defined by HM Treasury as 'uncertainty of outcome, whether positive opportunity or negative outcome'. However, others believe risk should not be confused with uncertainty, arguing the former is known about and an assessment of its probable impact made, whereas uncertainty is not known about and can have either a negative or positive effect. Clearly there is a conflict of views on this matter and the reader is well advised to seek clarification when reading others' discourses. There is a strong correlation between risk management and value engineering. The two subjects are linked as any value management judgement can alter the risk.

Risk management is concerned with identifying relevant risks, assessing their likelihood and impact, and deciding how best to manage them. It is not about avoiding risk, for to do so would remove any entrepreneurial spirit in a team and life would be come boring. Risk taking is part of normal business practice; what needs to be done is to take calculated risks. There is a difference between accepting risk and ignoring it. There is a risk involved in crossing the road, but by taking simple steps such as looking in both directions first (remember the green cross code) or better still using a pedestrian crossing, underpass, or bridge, the risks are reduced.

Is there a cost associated with managing risk? This is a difficult question to answer as it depends on circumstances. In the case of the pedestrian crossing the road, then yes, it costs to provide physical alternatives such as pedestrian crossings, but on the other hand what is the cost if a person is injured or killed if this provision is not available? If a client shifts the financial risk to the contractor, those that are vigilant will make allowance for this within their price for the works to counterbalance this disadvantage. Whether or not this makes the overall cost of the contract more expensive or cheaper is an interesting dilemma.

10.2 Dealing with risks

Having accepted that risks and risk taking are inherent to business, and construction is no exception, a procedure has to be adopted to deal with risk. The first stage is to identify the risks. Irrespective of the stage in the process it is necessary for the team to investigate the part of the process they are involved in and identify all potential risks. Further, they should look beyond their remit and consider all the activities prior to and after their area of responsibility that might affect them, such as maintenance issues and demolition when designing the new building. All of this in the context of the client's clearly stated priority of requirements. For example, these may be cost, programme, quality, prestigious design, impact on neighbours, environmentally sensitivity, or safety during the construction works.

10.3 Assessment of risks

Having established what the likely risks are, it is then possible to analyse what the implications are. It is difficult to quantify in a few words how to do this because the types of risks are so variable. For example, if it is a safety risk, it is possible to assess the frequency of an event occurring and the severity of the harm likely to result in the event of an accident occurring. This enables a risk assessment to be made (*Operations Management for Construction*, Chapter 4). Assessing the probability of an unforeseen circumstance in the ground is a singular occurrence and is more of a function of the quantity and quantity of site investigation research carried out. Financial risks associated with property development in obtaining a return on one's investment are notoriously difficult to predict (*Finance and Control for Construction*, Chapter 3) due to the many variables involved and the timescale of the investment. In any of these cases it is important to assess which risks are likely to have the greatest impact so that sufficient energies can be invested in establishing what they are and how to protect against them. Risk assessment is the systematic identification of risks and their relative risk compared against the others so management has a full and clear understanding of the risks faced.

10.4 Responding to risks

Having established the risk there are several ways to react to it. The most common in the development and contract process is to transfer a portion of the risk to others. For example, it would be unusual for a property developer of a significant project to take all the financial risk and much of the risk will be transferred to other financial organisations that take a value judgement as to the likely success or failure of the project. This is similar to the workings of Lloyds Insurance when the Names are offered the opportunity to take a proportion of the risk of insuring a specified situation, property or object. Laying off risk to others by necessity involves paying a premium to whoever takes on the risk. This is after all one of the ways they make their money. On assessing the risk, if they consider it to be small their premium will be lower than if it is considered to be high. Their premium will also be calculated on their knowledge of the client. In the case of insurance, the greater the track record of no claims, the lower the premium; and in the case of property development, the proven successful developer would expect to pay less.

Transfer of risk between the client and the contractor is achieved by using clauses in the contract between the two parties. An example of this would be in the JCT05 clauses on fluctuations. If the client requires a fixed price the contractor has to price with inflationary pressures in mind and if, through the pressures of competition, it underestimates, it will be taking the risk. On the other hand if the client uses a fluctuation clause the risk remains with the client rather than the contractor. In this case the risk is manageable as, unless there is some dramatic national or global economic incident, it is possible to make an informed decision about price. If the risk is more difficult to predict the contractor is more likely to add on a higher premium to cover for all eventualities. This clearly suggests that the client would be better not trying to pass on the risk as it could cost more in the long term, either because it will cost more or a contractor takes the risk, loses out and goes in to liquidation. It is reasonable, therefore, to accept risk as part of business responsibility rather than try to transfer it. When transferring risk it is essential the risk is understood, the most appropriate place is sought to manage it, and the cost of either keeping the risk or transferring it should be established.

An alternative to transfer of risk is to insure against risk. Homeowners will insure against risk for contents and buildings in case of an accident. The greater the protection required, the greater the premium. Likewise in contracting it is normal to have cover for indemnity against third-party claims (liability insurance) and fire. Professional indemnity insurance will be sought for those involved in the design process, including architects, engineers and quantity and building surveyors, to cover for mistakes they might make. Contractors offering design and build services will also have this cover.

Reduction and elimination is another way to deal with risk. This is sometimes referred to as mitigation and we practice this everyday, like carrying an umbrella in case it rains or having a mobile phone in case the car breaks down. Examples in construction include value engineering studies producing an alternative design solution; improving the accuracy of information upon which estimates and construction programmes can be prepared, such as more detailed design or further site investigation; changing the method of construction so as the reduce safety risks; and selecting an alternative procurement strategy so as to allocate risk between the various parties in a more appropriate way.

A further approach to risk is to avoid taking it. Once the risks have been identified, it may be decided that the risks are too great and no further progress will be made. For example, a property developer may consider the probability of the project making an appropriate financial return is not good enough. Alternatively, a contractor decides there are too many unknowns about the project and excessive risk has been transferred in the conditions of contract making the overall risk too great, so declines to tender.

The final approach to risk is to ignore it all together. In other words, bury one's head in the sand. Many in design and contracting still do not analyse the risk of the project they are embarking on, relying on the assumption the standard conditions of contract used will deal with risk, then wonder why matters go wrong. This is because the balance of risk may not be pertinent to the specific project and also the standard forms do not cover for every eventuality.

Figure 10.1 is a risk matrix adapted from the *Business Continuity Planning Guide* (1998) and is an interesting way of assisting in deciding how to deal with the different ways in dealing with risk.

Risk should be reviewed at least once a year, and more frequently if significant changes have occurred, be they internally or externally generated. A sudden rise in global oil prices might impact on a haulage business significantly. Counter-measures put in place may need to be checked, amended or removed all together if the threat has disappeared. The police are continually doing this in light of changing circumstances and threats.

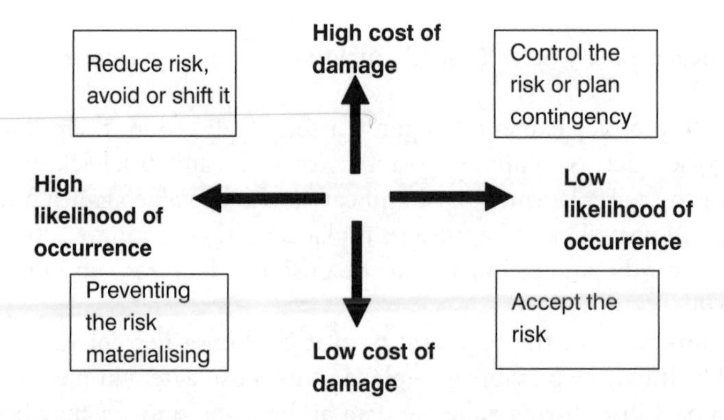

Figure 10.1 Risk matrix

10.5 Conducting a risk assessment

It would be difficult to assess risk across the whole of an organisation because of the scale of the task and its comprehension. It is therefore normal to break down the business into manageable and discrete activities, for example, production safety, computer security, development, finance and suppliers.

Those involved in risk assessment need to be experienced and qualified to carry out this task and have sufficient knowledge of the area of work being assessed so they know what questions to ask and how to interpret the answers given. The kinds of questions asked could include:

- What currently are the hazards, failures, breakdowns or interruptions facing the organisation?
- If any of these occurred, what would the effects be on the organisation, business partners, customers, our staff, supply chain, local community, environment, etc.?
- What is the likelihood of any of these hazards, breakdown, failure or interruptions occurring?
- Are we prepared for such an event?
- Is the risk acceptable to the organisation and others likely to be effected?
- Can the level of risk be controlled?
- Is there anything the organisation can do about it in terms of being prepared or taking steps for recovery?
- Are there any restraints that restrict preparation or recovery?

They will establish the facts by interviewing, physically inspecting the workplace and looking at data on previous incidents following a methodical sequence. As an example, an assessment looking into the risks associated with physical assets, would normally include the following stages:

- List all the department/company assets and group them into like categories such as computers, printers, scanners and disk backup storage.
- Establish the replacement cost and categorise its value against a table of ranges of cost. This is because to replace an item costing £100,000 has a significantly greater impact on the business than does an item valued at £100.
- Assess its value to the organisation if it could not be replaced easily or quickly. It may be possible to replace an expensive item immediately, but a compact disc storing valuable data and costing only £1 may be either irreplaceable or be very disruptive to the business if damaged, destroyed, lost or stolen.
- Consider the likely threats each asset is subjected to, such as fire, theft or breakdown, and for each of these threats, assess the likelihood of this occurring on a rating scale.
- Assess, if the threat materialises, how widespread an impact it will have on the business.
- Assess the overall risk either by a valued judgement or using calculations.
- Draw comparisons of all the risks to the assets.

10.6 Risk assessment calculations

The following is a typical example of one of the simpler approaches to give the reader a flavour of the issues. To obtain a further insight into more sophisticated methods see Chicken (1996). A matrix can be produced as shown in Table 10.1. The left-hand column shows the issues that might be assessed for a particular asset, the type and likelihood of a threat and its vulnerability in the event of the threat happening, sometimes referred to as

Table 10.1 Assessment of risk

	Asset A	*Asset B*	*Asset C*	*Asset D*	*Asset E*
Asset value (£)	2	4	1	5	2
Likelihood of threat	1	1	3	3	2
Vulnerability	3	4	1	2	5
Assessment of the risk	2.00	3.00	1.67	3.33	3.00

the frequency of likely occurrence. The top row is different assets of similar groupings. A risk rating appropriate to each is entered into the matrix. These range from very low significance to high significance and are given an appropriate rating scale number of 1 to 5. The example shown is fictitious, but is used to demonstrate how the assessment of the risk is calculated.

The bottom row shows that the assessment of risk is calculated by adding together the three values (asset value, likelihood of threat and vulnerability) and dividing by three to obtain an average. The greater the value of assessment of risk is, the higher the risk to the asset in the business. The weakness of this simple system is that it assumes the values attributed to these three issues are of equal importance relatively, which of course may not be the case. In this example the asset has been identified as a financial cost, but it could equally be a measure of the safety of the staff. From these data and calculations it can be demonstrated where the greater risks are in the organisation so that risks can be prioritised and appropriate action taken. Safety risk assessments are discussed in *Operations Management for Construction*, Chapter 4.

10.7 Monitoring

Throughout any project there should be an ongoing process of risk management reviews. It is suggested in Achieving Excellence in Construction Procurement Guide (2003) there are several key times when this should occur. These are:

- When the objectives and priorities of the main parties of a development project have been identified, a risk assessment of the potential project options should be made.
- Having decided upon a development project there are various options available which could meet the user's requirements. Once these alternatives have been considered it is necessary to carry out a detailed analysis looking at the balance cost and the risk of each solution before making the final decision.
- Selecting the type of procurement system best suited to the project, by applying a risk assessment of each of those available and considering how best to manage the risk in each case as outlined above (transfer, avoidance, reduction, acceptance or sharing).
- In the outline design process, the risk is associated with whole-life costing and life-cycle analysis and buildability aspects as errors here can have a significant long-term impact on the financial performance of the building during construction and later on during its life in terms of

running and maintenance costs, as well as the success or otherwise of its functional use.

- During the detailed design process, the risks are concerned with the selection of materials, especially finishes. These affect frequency of replacement or planned maintenance and ease or otherwise of cleaning. This process will be ongoing during the construction of the building.
- The final stage of risk management is concerned with facilities management. The decision on how this will be accomplished may have already been looked at the procurement process, such as the service contract management phase of a PFI project. Alternatively, the management of this process may be done in-house, in which case the risk assessment needs to be carried out to decide how much is carried out by the company's employees and how much is out-sourced and how these sub-contractors are controlled.

10.8 Feedback

This stage is about analysing how well risks were managed and what lessons can be learnt to improve risk management in the future. It is often overlooked as a result of new tasks and challenges, but should be encouraged.

An interesting example of acceptance (or otherwise) of risk is the debate at the Health and Safety Executive (HSE) which is no doubt frustrated at their inability to pin blame on a corporate body, and the bodies representing employers. The HSE wishes a named director to be responsible for safety and the employers are opposed to this, no doubt as they consider it is less onerous to hide behind corporate responsibility, thereby reducing the risk of a board member being jailed in the case of a serious accident.

10.9 When do risks occur throughout the process?

It is important to realise that risks associated with the development and construction of building runs through until the building is finally demolished and disposed of.

10.9.1 Development viability and feasibility stages

When developers see an opportunity for a possible development, they look at the expected return on their investment for a particular use of the project on completion be that from rent or sales, and look at different sources of funding. The risks involved in this process include:

- deciding upon the use of the site;
- obtaining planning permission;
- estimating the costs of the construction works when the design is not complete so it is difficult to assess accurately and relies instead on ballpark figures – there are uncertainties about the ground conditions, decontamination costs, archaeological finds;
- obtaining the completed development on time;
- estimating the costs of facilities management;
- selection of the team to provide the information to enact the above;
- deciding on whether to sell or lease;
- estimating the costs involved in selling or leasing and if the latter, estimating the length of time to lease all the property;
- obtaining the returns estimated;
- obtaining the finance at the right interest rates and the implications if there are fluctuations;
- accuracy of calculations: comparative, residual, and discounted cash flow methods.

Factors affecting the calculations are primarily economic and include the problems associated with the many variables in predicting interest rate changes, wage inflation, increase in the cost of materials and the state of the stock market. Also, clients would not wish to commit finance to a project until all the risks have been identified and assessed and agreed how they are to be managed.

10.9.2 Design stage

Much of the risk at this stage revolves around the strength of the relationship between the client and the design team and within the design group itself. The likely risks include:

- the lack of understanding of the client's needs by the design team;
- whether or not the client really understands what is needed – often the client has a view of what is needed, but it is different from that of the people who have to live and work in the building.
- poor briefings by the client to the designers which of course the designers have a responsibility in rectifying.
- designing solutions that are complex, difficult, extend the contract duration and hence are costly to produce, i.e. buildability issues.
- designing the risk of crime out of buildings or the complex if more than one building, such as a housing estate.

- breakdowns in communications between the design team partners;
- inadequate research into ground conditions;
- the use of inappropriate technologies;
- designing buildings that are difficult or expensive to maintain;
- inaccurate cost planning;
- overshooting the budget;
- satisfying their own needs in producing an 'architectural masterpiece'.

Value engineering at this stage can have a significant effect on improving the overall quality of the building performance and reducing risk (*Finance and Control for Construction,* Chapter7).

10.9.3 Procurement stage

It is important that the selected method of procurement suits the type and needs of the project (*Finance and Control for Construction,* Chapter 8). Much of the risk depends upon the priority to which the client gives to the following:

- their experience and how much they wish to be involved in the construction process;
- separation of the design process from the management of the construction process;
- the need to alter the specification and design as the project develops building in a certain amount of flexibility into the briefing process;
- the ability to seek for remedies if dissatisfied with the design and construction processes;
- the complexity of the project;
- the speed at which the contract needs to be completed from inception to handover;
- how much flexibility there is from the budget price to the final price.

There is also risk to the contractor depending upon the procurement rate, which is in essence often the reciprocal to that of the client.

10.9.4 Tendering stage

The following risks are inherent in the tendering process itself:

- the amount of time predicted to be lost as a result of inclement weather is notoriously difficult to assess;

- in the euphoria of wanting to successfully obtain the contract and meet the client's needs there is a danger of making optimistic assumptions when programming;
- the time available for carrying out the estimate is limited so there is always the risk that not enough time is allocated to thinking through all the implications of the construction processes and the wrong method of works is selected as the basis for the tender;
- deciding on the levels of overheads and profit to be added is the greatest risk of all, which is why senior management takes the decision;
- predicting the increased costs of plant, materials and labour throughout the duration of the project, especially if a long contract is difficult – the risk is reduced in the event of having a fluctuation price contract;
- if during the course of the project shortages of labour, plant and materials occur, the price of obtaining them will rise;
- in the end, other than when insufficient overheads and profits have been added, the contracts success or failure is usually a reflection of the quality of the management.

It is interesting that many of the risks can be reduced significantly if all parties work with a spirit of co-operation rather than confrontation.

10.9.5 Construction stage

The first thought concerning risk in the construction process is naturally that of health and safety (*Operations Management for Construction,* Chapter 4). However there are other risks associated with this stage, some of which will have been assessed at the tendering stage. These include:

- sensitive projects can attract objectors who may take positive action to prevent construction work from occurring – notable examples in recent years have been motorways and bypasses, such as Newbury and the second runway at Manchester airport.
- theft from the general public, the contractor's employees and organised criminals.
- terrorist activity on high-profile buildings works – this is not just about attacks on the building works, but implanting devices to set off later.
- vandalism.
- trespassers and possible injury incurred by them.
- fire, which may be started by accident or deliberately by vandals.
- inclement weather, which will almost certainly have been taken account of in the estimate, but is unpredictable;

- industrial action is less likely than in the 1960s but is still a possibility, not necessarily always on the site but also to a supplier;
- shortages of labour and materials;
- unforeseen construction problems;
- design deficiencies;
- delays in information being provided;
- complaints from neighbours about the nuisance value of the construction work.;
- spillages of pollutants either into the ground on entering water courses and sewage systems;
- disposal of hazardous materials in general.

10.9.6 Commissioning and snagging stage

The amount of risk in the commissioning stage is a function of the complexity of the building. Clearly the greater the percentage of services in the project, the more it is likely to go wrong and this needs to be taken account of in the planning of the commissioning and installing processes. It is important to build in adequate time for testing and rectifying any faults found. How much snagging is involved is directly related to the methods in which quality is controlled and assured during the construction process.

10.9.7 Occupation of the building

The main question here is what is the impact on the business of the building users if something happens. This is developed further under business continuity development (section 10.10). Typical examples include:

- computer failure caused by damage of some kind such as fire, theft of the machine or information and computer viruses;
- fire destroying part or all of the premises;
- burglary, theft by personnel and vandalism;
- disclosure of information by company staff to others outside of the organisation – sometimes for financial gain and in others for political or mischievous purposes often referred to as 'leaks'.;
- assault on personnel in or near to the building;
- threats from terrorists;
- radiation contamination, which could occur in hospitals and science laboratories;
- the accuracy of the planned maintenance programme;

- the speed of response to maintenance and replacement requests, e.g. light bulbs in an operating theatre during an operation;
- sufficient retention monies withheld from the contractor to cover remedial work;
- whether changes of use be accommodated;
- whether the budgets for facilities management accurate.

10.9.8 Demolition stage

- safety issues are of paramount importance and potentially are the greatest risk – these include not just the damages of collapse during the process, but also dusts, fibres and fumes resulting from previous use of the building and of course the discovery of asbestos;
- disposal of the materials comes with its own risk – materials need to be identified so they can be safely disposed of or recycled;
- the financial risk has to do with having a methodology of demolition that can be carried out without short cuts having to be taken;
- assumptions as the amount of material that can be recycled;
- difficulty of dismantling;
- dangers to third parties, especially if using explosives;
- pollution of the atmosphere and water courses during the process.

10.10 Business continuity management (BCM)

'Planning which identifies an organisation's exposure to internal and external threats and synthesises hard and soft assets to provide effective prevention and recovery for an organisation, whilst maintaining competitive advantage and value system integrity.' (Herbane et al. 1997)

There has been an increasing awareness of the sorts of issues raised above, in the occupation of the building, of the need to have a business plan in place that can be brought into action as a result of a serious disruption to a business. Notable examples reminding of the need, being the Saturday, 15 June 1996 bomb blast in the centre of Manchester and the attack on the World Trade Center in New York on 11 September 2001. One of the buildings severely damaged in the Manchester blast, later demolished, housed the offices of the Royal Sun Alliance. They had a recovery plan in place and found alternative accommodation by the following Monday. They also found a warehouse to deal with recovering data from damaged computers and the documents taken from the damaged building although in reality it was several days before access

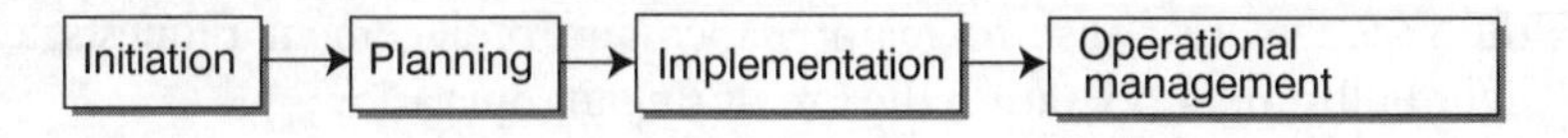

Figure 10.2 The continuity process

was permitted to recover the information whilst a structural assessment of the building was made. However, not all interruptions to business are as dramatic as these two examples, but can be as equally damaging as can be seen when the building housing the business is burnt to the ground. Many interruptions are less severe, but still cause many problems for all those concerned.

The concept of business continuity management has been a developing process much of which is based on the previous discussions in this chapter. Initially it was seen as dealing solely with the physical equipment, primarily computing, and was concerned with answering the question 'what would happen in the event of a technological failure?' This developed into first about thinking about attacks against the system and the need to provide security to prevent these attacks from being successful, and second to implement procedures in the event of a successful attack, such as back-up systems and the development of a survival plan to ensure survival and recovery.

Companies realised that recovery is not just about the physical assets of the business, but that there are people involved in this process who are the most important resource the company utilises. Take the example of Royal Sun Alliance in Manchester: they found alternative office space in their Liverpool offices as well as finding space elsewhere in Manchester, but what about if their staff are unable to work because of their domestic needs or because of any residual trauma? For example, there is no way a part-time working, single parent with children at school, previously working in the Manchester office, can readily commute from Manchester to Liverpool.

There are four stages in the continuity management process as shown Figure 10.2.

10.9.1 Initiation

Initiation involves changing the behaviour of the managers and employees in the organisation. As in all significant changes within an organisation, whether it is in behavioural, procedural or process, the drive must come from the very top. Management must identify and determine what parts of the business are likely to be needed to be covered by the continuity plan, whether it can be developed internally or with outside assistance, and then set the policy. The likely impact on the business is part of the continuity planning process. These policy decisions need to then be clearly communicated to all stakeholders in the business. As with any policy, structures within the organisation need to be created and key staff selected and trained so the policy can be implemented. In other words, it is necessary to have enthusiasts at each appropriate level. Sufficient finance also needs to be set aside for the initial setting up of the plan, its introduction and maintenance. How much is allocated depends on what issues have been identified. In carrying out this work it is imperative that the key staff are able to convince others in the organisation of the need, and one of the ways of doing this is to create scenarios that might happen and ask the question: 'What would the impact be on them and the company if this event took place?'

10.9.2 Continuity planning

The plan is a guide for action in the event of an interruption of part or all of the organisation's business. It is necessary to identify a list of crises that might affect the business. Some of the specific issues related to the development and construction processes have already been identified, but Table 10.2, modified from Elliott *et al.* (2002) gives a more general view of likely hazards.

Clearly not all will be relevant to the business of the organisation. The objective of the exercise is to establish, primarily, the likelihood and the impact such an event would have on the business. The timing of an event should be considered, as the work in some businesses will not be the same throughout the year. For example, what would happen if the system for communicating national exam results to school children broke down.

On identifying likely causes an analysis of the impact of such an event has to be established. So, for example, if there was an electrical power failure or fault, a whole range of effects could be expected besides the loss of

Table 10.2 Likely hazards

Adverse weather	Hostile takeover	Plant failure (e.g. tower crane collapse)
Computer breakdown	Illegal activities	Product tampering
Computer viruses	Industrial action	Sabotage by outsiders
Computer failures	Kidnapping	Sabotage by staff
Currency fluctuations	Loss of important staff	Supplier/sub-contractor goes bankrupt
Design failure	Major industrial accidents	Telecommunications failure
Disease/epidemic	Natural disasters	Terrorism
Fire	Negative publicity	
Floods		

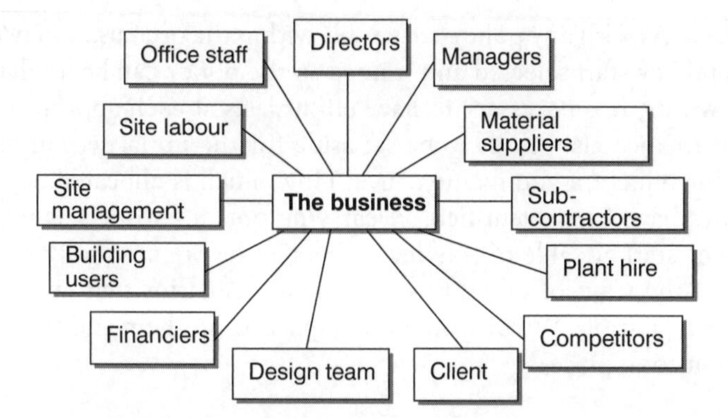

Figure 10.3 Main stakeholders to a construction company

power, such as power surges and loss of voltage which may affect sensitive electrically driven equipment and machinery. This is referred to as business impact analysis and whilst not within the remit of this text to demonstrate methods of assessing the impact (See Elliott *et al.* 2002), it is important to realise who may be affected by an event.

Businesses do not work in isolation, as there are many stakeholders involved, both within the organisation and externally. Typically in a construction company they are as demonstrated in Figure 10.3.

Impact caused by interruption is a complex process. A breakdown caused by one supplier, does not just impact on the contractor. If the contract is delayed, besides possible penalties resulting from a late handover, the client's business is affected, as are some or all of the other sub-contractors and suppliers. Reputation may be damaged which have an effect in obtaining work in the future thereby introducing the possibility of redundancies within the company. A possible overstated bleak scenario, but is it? As supply chain management develops, it will be in the contractor's interest to assist those in the supply chain with their own business continuity plans as many of them will not have the necessary resources to carry it out themselves.

There are others who may be involved in some way as a result of an incident. The general public is affected by noise and pollution and in some cases is injured. The media report incidents that can in some cases be damaging to the reputation of the business and hence the flow of information to the media needs to be well controlled (Chapter 5). Others who can impact on the business include protesters, trade unions and their members, regulatory bodies and governments.

After the analysis of the likely causes of interruptions, the next stage is to conduct a business impact evaluation. This involves setting objectives to recover from the interruption. For example, if the school is burnt down, the activity within it must be up and running elsewhere within one week. If the tower crane collapses, the site must be operational within one working week (subject to HSE approval). If the telephones fail, alternative communication must be provided within 30 minutes.

Structured approach risk assessments have to be carried out to identify the likelihood of interruptions occurring and the consequences if they do from which the priorities for a business recovery plan can be established. The final stage is to develop business interruption scenarios, i.e. what happens if, and how do we deal with it if it occurs and decide how often these should be tested? This works well if there is wide participation within the organisation, bringing together all the inherent skill and expertise. This acts as the basis of producing the business continuity plan.

10.9.3 Implementation

In many senses producing the continuity plan is the easy part. Implementation of anything new requires skill and competent management. It comprises four stages, that of communicating the reasons for the plan and the implications to the individuals and the organisation as a whole; developing the culture and business ethic within the organisation to cope with the changes; introducing new control systems; and training the personal to carry out new tasks.

10.9.4 Operational management

Having established the likely risks and introduced systems of management and control into the organisation, will they work when an incident occurs? There are several stages to consider. How much time and effort is spent on each stage depends upon the likelihood and the magnitude of the effect if an incident occurs.

The first stage is to put together teams of staff who will be responsible for dealing with the incident, often referred to as the crisis management team. Their responsibility is to organise the stages mentioned below and deal with incidents in general. How well they operate is often a function of the type of culture that exists in the organisation as a whole and the ability of the business to create teams and foster team building. This involves full participation and contribution from all members, brainstorming sessions, complete openness and employing outside experts where necessary.

The second stage is to put in place a command and control structure that will be able to cope with incidents. This is important so that consistency of approach can be adopted in each case. This does not mean to say all levels of the command structure will be used for each incident as this depends upon its severity and impact on the business. The more serious the problem, the higher up the chain of command the control will be. Specific personnel need to have clearly defined roles such as who will deal with the media and the emergency services. For example if a fire occurs in a large sprawling industrial complex, it is no good a member of staff contacting the fire brigade saying there is a fire at the works. The fire services will need clear instructions as to where the fire is and how to get there as quickly as possible. This may be advising security of the fire who will then liaise with the brigade.

The third stage is testing and auditing. A clear example of this are the rehearsals the emergency services conduct from time to time on the more dramatic incidents such as dealing with a 'weapons of mass destruction' incident or an aeroplane crash. The purpose of testing is to find out whether the systems in place will work, staff have the right attitude and have not become complacent, and to maintain awareness throughout the business. The frequency of tests will have been determined when planning. The types of test used are as follows:

- *Desk check tests.* These are carried out regularly and frequently and involve checking that all the information contained in the plan is correct and up to date, such as named persons and telephone numbers. Also, that the plan still matches the department/organisation's current work as this may have changed from when the last plan was introduced.
- *Walk-though tests.* All the named persons are brought together and asked to role-play their procedures. This highlights any communication difficulties and whether the activities are being completed within the prescribed times and sequence.
- *Simulation exercises.* Advance notice is given to all those concerned. A scenario is then created which may restrict the participants in their use of standard communication methods or to parts of the building. Under observation they then have to set in motion the recovery plan and execute it to completion. The time allowed may be accelerated from that allowed for in a real incident. The observers look for breakdowns in the participants' performance either caused by their errors or in the plan itself. A revised plan can then be introduced after the exercise and subsequent debrief.
- *Operational tests.* Here a complete department will be closed down and the employees have to relocate in another building either owned by the organisation or supplied by others and restart the business activity within a prescribed time. This tests the ability of the staff to carry out the exercise and if there is enough backup resource to permit it to work.

- *Live exercises.* The difference between this type of test and the rest is that no prior notice is given. The simplest type would be evacuating the building in a fire drill, but could range up to dealing with a significant interruption of the work. These tests can raise the awareness of the impact on all of the organisation's employees. They can of course be expensive to enact.

The debrief is imperative in these tests. The objective is to establish how successful the continuity plans are in practice, and how they should be modified for the future. The observers should also question their role to see if they could improve their performance.

It should also be remembered that in the case of simulation, operational and live tests, the participants can suffer high levels of stress as they become more actively engaged in the process and attempt to put the plans into operation which rarely go strictly to plan. These stress levels should be noted, as it may be necessary as a result of the findings to introduce training for staff and more important educate management to manage stress in others better.

References

Chicken, J. (1996) *Risk Handbook.* International Thomson Business Press.

Elliott, D., Swartz, E. and Herbane, B. (2002) *Business Continuity Management.* Routledge.

Herbane B., Elliott, D. and Swartz. E. (1997) Contingency and continua: achieving excellence through business continuity planning. *Business Horizons*, 40(6): 19–25.

Kelly, J., Morledge, R. and Wilkinson, S. (2002) *Best Value in Construction.* Blackwell Puiblishing.

Murdoch, J. and Hughes. W. (2000) *Construction Contracts: Law and Management*, 3rd edn. E&FN Spon.

Office of Government Commerce (2003) *Achieving Excellence in Construction, Procurement Guide 04: Risk and Value Management.* OGC.

Property Advisors to the Civil Estate (1998) *Business Continuity Planning Guide.* OGC.

CHAPTER

Communications

> Now what is the message there? The message is that there are known knowns. There are things we know that we know. There are known unknowns. That is to say, there are things that we now know we don't know. But there are also unknown unknowns. There are things we don't know we don't know. So when we do the best we can and we pull all this information together, and we then say well that's basically what we see as the situation, that is really only the known knowns and the known unknowns. And each year, we discover a few more of those unknown unknowns.
>
> Donald Rumsfeld, US Secretary of Defense, June 2006.

11.1 Introduction

The scope of the subject is immense and to do it justice requires much more space than is available here, so the author has produced mainly bullet points to assist the reader in understanding the topic and hopefully give some indicators about the different aspects of the subject. To assist in understanding, Figure 11.1 visualises the various components to be considered. It is not a definitive solution but hopefully will go someway to help.

11.2 Personal communication

This is divided into three categories for simplicity: one to one, in a group, public speaking or presenting to an audience.

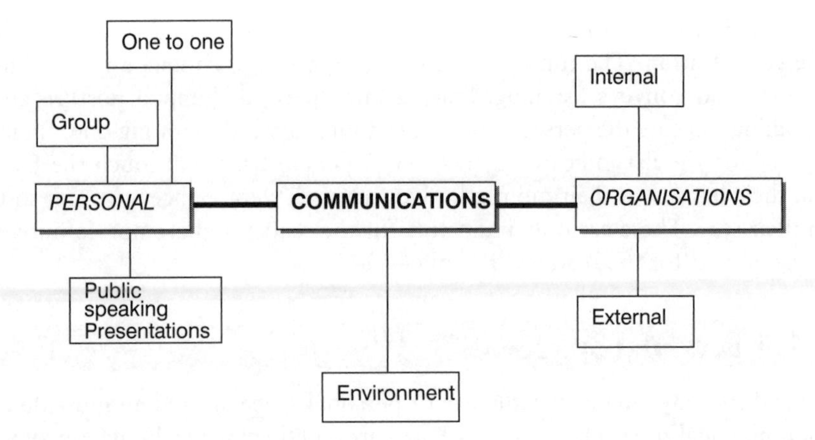

Figure 11.1 Communications

Within these three groups there are inevitably overlaps in some of the topics discussed whilst others are specific to the categories. Likewise since people work in organisations, many of the topics are relevant to organisations. However, for the purpose of this text much of the discussion under organisations will be confined to systems of communication.

11.3 Body language

One of the fundamental ways of communicating is using body language. For example, dogs wag their tails, snarl or roll over on their backs to indicate to others happiness, fear/warning, and submission respectively. Throughout the animal kingdom there are many, far to numerous to mention here, examples of the use of body language. Humans are no different except we have become less tuned to reading the signals.

We have lost the knack of interpretation as the human race has developed other highly sophisticated methods of communication, notably speech and writing. However, there are others such as music and painting, which equally have a place in the field of communications. Architecture also should be mentioned, as it is often a statement of the mood of society at a point in time. For example, building estates with high fences or walls around is commonplace at the moment, clearly reflecting our current society's values and fears. For those who have lost a major sensory power such as speech or hearing, signing has been developed as means of communication.

There are a variety of ways at looking at body language. First, there are the emotions conveyed by facial expressions: a smile, a grin, a scowl, a frown. Then there are the physical movements of the body often conveyed subconsciously. A person leaning forward with open arms is engaged in the conversation. The tilting of the head, a lot of eye contact and nodding of the head conveys listening. Staring into space, a slumped posture and doodling indicate the person is bored. Leaning forward, pointing ones finger or shaking the fist indicates aggression. A person lying will touch the face, put their hand over their mouth, look down to the left, glance at you or shift in their seat. These are only indicators for the reader and are not definitive. A useful text for the reader is Kuhnke(2007).

11.4 Personal space

Linked to body language is the use of personal space around an individual: their personal zone. This zone is the area around them which humans protect from the intrusion of outsiders and can have a direct effect on the way they do or do not communicate.

The following distances act as a guide to define the acceptable spaces for various modes of communication:

- the intimate distance for embracing or whispering (0.15–0.45m);
- the personal distance for conversations among good friends (0.45–1.2m);
- the social distance for conversations among acquaintances (1.2m–4.0m);
- the public distance used for public speaking (4m or more).

These distances may vary between cultures. However, what matters is the comprehension there are personal zones, which if invaded, makes a person uncomfortable and thus likely to react differently from normal because of increased agitation. Police interrogators are taught that violation of personal space can offer a psychological advantage accomplished by sitting close and crowding the suspect.

There are also territorial areas within the office environment. This is not surprising. Animals not only identify their territory, but also in many cases vigorously defend it, sometimes to the death, but preferably by making the invader climb down and retreat.

Humans are no different. Take the classroom scenario. How many students always sit in the same seat in the classroom and how do they feel when someone else 'pinches' their seat, even though it doesn't actually belong to them? Regulars in pubs start to commandeer their space at the bar or their seat at a table. It happens in church where the majority of parishioners sit in the same pew each week.

11.5 Environment

Before discussing other types of communication it is important to consider the environment in which communication is taking place as these issues can affect the quality and effectiveness of the communication.These include noise, colour, atmospheric conditions, seating and fear.

The volume and type of noise has a profound effect on the ability to communicate. Too loud, as at a disco, makes conversation difficult and frustrating. Regular very loud noise also disrupts the ability to listen. For example, many of the schools in the Kowloon peninsula in Hong Kong were under the flight path into the Kai Tak airport. Large passenger jets flew in every two or three minutes throughout the working day often peaking at 105db or more. As the aeroplane approached the children would put their hands over their ears long before the pain barrier was reached and keep them there until the level of sound had reduced. The impact on the learning process was immense.

The type of sound also has an effect. Background continuous sound, as often occurs on site, can be tolerated. Unexpected irritating or loud 'bangs' are more difficult to accept. Some use background music as a method of drowning out other noises, but again the type of music will vary by individual.

The décor, especially colours, can affect the way in which we communicate. Most offices are decorated in light pastel shades, all of which convey calm and serenity and are, literally, easy on the eyes. Bright or harsh colours can have exactly the opposite effect and cause agitation and discomfort.

Temperature and quality of the air both have an effect on the ability to work and therefore concentrate. If the space is too hot most people become lethargic and sleepy, unable to fully concentrate, listen and think – all pre-requisites of communication. Equally, if cold, people are more concerned with keeping warm than concentrating on the task in hand. Air pollution can also have an effect. Smelly environments to a varying degree cause distraction, as do high levels of dust or fumes.

There is seating which is appropriate to a task, and seating that is not. If too comfortable it becomes increasingly difficulty to concentrate, similarly if uncomfortable. Airline seats are notorious for being difficult to sleep in and how many people complain of backache when working for long periods at a computer station, usually caused by either not having the appropriate seat, or if provided, not adjusting it correctly.

The relationship between levels of management and colleagues is important to achieve good communications. If an atmosphere of fear is the norm in the workplace, then it is very difficult to communicate properly. On the basis that communication is usually a two-way process, fear restricts this from happening.

11.6 One-to-one communication

This may involve any one or combination of conversations: face to face, over the telephone, correspondence, and sketching and drawing.

11.6.1 Conversations face to face

This can be split into further subcategories: general conversation, giving or receiving of orders, debating, negotiating, interviewing and disciplining. In all of these cases the same basic principles should be applied. Often conversations drift along, which is usually a waste of time. Not always though, as sometimes chatter is a useful way to break the ice and get to know the other person. However, generally it is unproductive. The first priority is to have your message understood by the other party. Second, one must receive and understand the message sent. Third, neither of these will be achieved if one does not develop the ability to listen and observe. Listening is crucial to this as was identified in section 8.8.8.

A good example of this and what might happen if it goes wrong is:

> What was said was: 'Check the edge of the formwork and reinforce it.'
>
> What was heard was: 'Check the edge of the formwork and reinforcement.'

So, what steps can be taken to ensure that the message is understood?

Avoid ambiguity. Ambiguity can have two effects, the more serious being the receiver of the instruction goes and does the wrong thing, or alternatively, asks for clarification, which wastes time. How do you understand the following: this country is dry, it was a funny meeting, a couple minutes, a sustainable business, or indeed, the term communication. Vocabulary should be understandable. Technical jargon used only by the specific business or organisation is alien to others: for example, formwork, water–cement ratio, planning supervisor (which is also ambiguous – safety or programming or town and country planning). Short hand, especially acronyms such as HSE, and AA (also ambiguous!) can also be confusing. Rarely used words which are not in common use and local dialect and slang all can add to confusion.

Use relatively short sentences. This clearly depends on the ability of the other party to receive information verbally, but if the sentence is too long it is easy to lose track of the message.

Consider the speed of delivery. Again this depends on the competence of the individual to comprehend what is being said. For example, someone with limited knowledge of English would normally have to be spoken to more slowly than a person who is fluent in English. Equally, if the concepts or the technical content are complex it might be necessary to deliver more slowly, and to repeat the message to reinforce the point.

Conciseness. Do not waffle. Get to the point as soon as possible taking account of the items above. However, depending on the reason for the conversation and who the third party is, it may be necessary to dwell for a short time on pleasantries to ensure they are at ease and 'switched on' or in tune with the other person. To give credence to this point, consider the start of a television programme. Generally nothing happens of significance for up to a minute to allow the viewers to mentally adjust to the topic and/or to make themselves comfortable.

State purpose and draw conversation to a conclusion. It is necessary to ensure the other party is aware of the purpose of the conversation or, if they are asking to talk to you, that you are aware of the reason for the conversation. Equally the conversation should be brought to a timely and conclusive finish.

Keep the moral high ground. Whilst it is reasonable to be assertive when appropriate, and at times it is necessary to be confrontational, it is important to keep the moral high ground even though the other person may be stooping to less pleasant practices. It is important to be fair and consistent. Insults are ineffective and attempting to make the other lose face, especially in front of others, is unacceptable. Losing one's temper does not help, primarily because you will lose your train of thought. (There are occasionally times when controlled loss of temper for effect can be utilised, but even this is debatable and a dangerous route to follow.) Perhaps the adage 'management is not a popularity contest' should be extended to include, 'but is about receiving and maintaining respect'.

11.6.2 Conversations via the telephone

These have advantages and disadvantages based on the fact the person at the other end of the phone cannot be seen. The rules on conversations as outlined above still apply. However, there are several other things to consider depending on whether you know the person on the other end of the line well or not at all. There is also a danger of allowing the caller to queue jump other business. Issues include:

- Some people are nervous about using the telephone, although this is not as much of a problem as it used to be.
- The image of the person in terms of age and appearance is absent.

- The voice often gives a completely different impression about the person as to who they are and what they look like and vice versa.
- The position of the person in the organisation is not always apparent especially when the call is to a name rather than a position. Often in face-to-face conversations there is an exchange of calling cards or introductions are made.
- There can be time delays in responses on some overseas connections.
- The body language element is absent, so their reaction is unseen which can be a problem as often the body language gives an indication the person does not comprehend or is uncomfortable in some way.
- It is easier to be assertive and stronger willed if confident in using the phone.
- There is the added cost of using the phone on external calls as well as the costs of the time of the participants.
- Telephone manner is probably more relevant than in the face-to-face conversations as a result of the absence of body language.

11.6.3 Correspondence

The English language is second to none in giving users the ability to be imaginative, creative, informative and descriptive. Hence the depth of literature produced over the ages from so many great talents. However, business letters are not expected to be of such stature although sometimes little gems do appear.

There are two basic considerations: content and appearance. Appearance is extremely important. The receiver of the letter obtains an immediate impression of the sender, by the way the text on the paper looks. Does it look neat and tidy, is the font size and choice appropriate. Remember tastes in styles change over time. Then on reading, is it grammatically correct and are there any typing errors or spelling mistakes. As a result of desktop publishing, the business community is used to high-quality communications.

Content is a more complex topic and is divided into two sections: basic layout and the selection of words and style.

Basic layout will normally include:

- return address of the writer of the letter;
- return telephone, fax and email address;
- date of sending;
- file reference – this can be at the end of the letter;
- recipient's name and address;

- attention line – this is not always used, but is helpful where writing to a position, rather than a named person, say, 'for the attention of the managing director' – usually used with the salutation line 'Dear Sirs';
- salutation – Dear …;
- introduction/reference – this is referring to previous letter(s), its date and its reference, if there was one, and a short statement of the subject and reason for writing, perhaps just having the topic in bold; in any case the purpose is to draw the reader's attention to the subject and reason for writing;
- the main text explaining the subject fully and clearly without unnecessary details;
- the closing, which involves ending the letter by saying something that is helpful and/or courteous;
- complimentary close, such as Yours sincerely or faithfully;
- signature;
- the sender's name typed;
- enclosures – either just the word indicating there is some accompanying paperwork or a list of what is included;
- carbon copies or cc. – these are the names of the persons to whom copies of the letter has also been sent;
- blind carbon copies or bcc. – not shown on the letter, but on the other copies, so the receiver of the letter does not know who else has had copies.

Content issues include:

- the reader is probably very busy so wants to absorb the contents as rapidly as possible;
- think who the recipient is and the level of technical expertise or understanding they have which can affect the selection of words, this will also determine the tone of the letter;
- avoid jargon unless the recipient is in the same field – the same applies to acronyms;
- get to the point early – there is no need to use unnecessary adjectives;
- keep sentences and paragraphs as short as possible, although the occasional longer sentence breaks up the monotony;
- try to be pleasant even in letters of complaint (keep the moral high ground);
- use active verbs, for example, 'I have made a decision'. is stronger than, 'A decision has been reached by me';
- only use the first person if it is appropriate to do so;

- round off large figures unless it is a contractual letter or a quotation – for example, just over £2 million is better than £2,065,876
- use positive expressions rather than negatives – for example, 'I forgot' rather than, 'I did not remember'.
- finally, check it before sending it.

11.6.4 Email

The use of email has shown up some interesting changes in the way we communicate. It is rapid, and permits extensive information to be sent without incurring the costs of sending hard copies or indeed, assuming the receiver is 'environmentally' friendly, the need for any hard copy. The major problem with email is for the receiver. The danger is information overload caused by inconsiderate senders.

There also seems an acceptance that typos and misspellings are acceptable in a way they are not in letters. How long this will remain is still to be seen. When fax messages were originally sent they were often hand-written, but now the norm is to type to a high level of presentation.

- *Mailing lists.* Once your name is on a mailing list, it can be rather like receiving junk mail through the letterbox. The only difference being that with junk mail you can take one look at the envelope and 'recycle it'. The mailing list has to be opened just in case and this can take up precious time that could be better used elsewhere.
- *The office comic.* Those infuriating colleagues who seem to have nothing else to do but email jokes, messages, etc. of no interest to anybody but themselves and other like-minded people.
- *The socialite.* Seems determined to involve everyone in extra-mural activities after office hours. This person is of kind heart and their enthusiasm should not be dented. It is therefore difficult to control them.
- *Attachments.* Whilst the use of attachments can be very useful, if the sender has to pay for the time connected to the net, unedited lengthy documents can be unwanted.
- *Queue jumpers.* Those senders who send a message marked 'urgent', when in truth it is not.
- *Spam.* Other than having some form of intercept, this will remain problematic.

There has become a habit, as with phone text messages, to develop a new language, for example, 'U' in lieu of 'you' and '4' in lieu of 'for'. Whether or not this will be acceptable in business remains to be seen.

Finally, the sending of viruses is a never-ending problem. It is not suggested that the legitimate business user deliberately does this, however, the sender can be responsible for causing untold damage to third parties.

11.6.5 Fax messages

When first introduced, this was seen as a very rapid means of communicating complex information, contractual letters and instructions. In construction for example, a design problem could be discussed over the phone and the architect immediately make a rough sketch and send it by fax to the site. Much of the information sent in these early days was hand-written, but has been superseded because of the ability to scan pictures and send them electronically.

One small thought, the equipment used should be considered from an environmental standpoint. There are some machines that memorise the document, eject the original and then after sending the information, promptly produce a copy of what was sent.

11.6.6 Sketching and drawing

In construction, drawing and sketching are a significant method of communicating information. In one-to-one contact the majority of this will be by means of sketching. This can vary from quite sophisticated to jottings on a napkin or drawing on a wall with a piece of stone. Whilst it is not the intention to discuss how to draw and sketch, what is important is to realise it is a very effective way of communicating: 'a picture is worth a thousand words'.

11.6.7 The use of body movement (gesticulation)

This should not to be confused with body language. There is a tendency in the more conservative parts of our society to feel uncomfortable about exaggerated body movement, for example vigorous use of arm movements. Impersonators often pick on this type of expression when conveying their interpretation of a well-known personality. Yet not only can this use of the body convey extra meaning to a point, in some cases it is much simpler to describe something to the receiver. Try describing a spiral staircase using words rather than a movement of the hand and arm.

11.6.8 Interviewing

This is discussed in section 8.8.

11.7 Group communication

Clearly much of the discussion in the previous sections 11.3–11.6 are relevant to group communications and it is not the intention to duplicate unless a particular point needs to be stressed. The main areas of group activity are formal and informal meetings. These can be held around a desk, on site, via telephone meetings and video-conferencing.

11.7.1 Formal meetings

For the purposes of this discourse, formal meetings are defined as a group of people collected together to resolve an issue or issues, a formal circulated agenda, minutes taken, and various actions noted. These meetings will normally be part of the normal mechanism of running the business. Typical meetings of this type include the annual general meeting, board of directors' meetings, safety committee and site meetings.

11.7.2 The agenda

The agenda is the plan and structure of the meeting. It should be circulated with any accompanying papers well before the meeting, at least one week before. This gives the attendee the opportunity to read, digest and carry out any preparatory work they wish.

The agenda should:

- state where and when the meeting is to take place;
- indicate the order of business to be discussed;
- provide notice to any individual member of input specifically required from them;
- ideally indicate the amount of time it is proposed to spend on each agenda item, this indicates to the member the depth of discussion or seriousness of the item;
- ensure that each item is stated clearly.
- items might include:
  - apologies for absence
  - antroduction of members

    - chairperson's remarks, usually to set the tone and reason for the meeting
    - checking accuracy of the minutes from the previous meeting
    - matters arising and actions that have been taken or that are not covered in the current agenda
- the items of business – these should be ordered in terms of priority, for example:
    - urgent items
    - those needing careful scrutiny
    - less important items
    - the positioning of controversial items should be considered tactically
    - items should be grouped into topic areas if possible
- any other business;
- date and time of next meeting.

It may be necessary to have an accompanying letter to explain what the reason for the meeting is and what are the necessary outcomes.

11.7.3 Chairmanship

Whilst not intending to write on the full aspects of good and bad chairmanship, a brief resumé here should assist the reader to understand what is expected and some of the pitfalls. It is worthwhile sitting back and considering why is the meeting taking place. The answer to that may well be it is not necessary and it has just become the norm to have a meeting with this particular group every Monday. Having decided the meeting is appropriate then:

- The person controlling the meeting ideally should be able to stand back from the issues being discussed, control and direct the discussions and bring each agenda item to a satisfactory conclusion.
- This can be difficult if that person is one of the main contributors to the meeting, although often this is unavoidable.
- The meeting should be planned in advance.
- Remember everybody's time is valuable.
- Does everybody asked really have to attend?
- Ensure all members are aware of what the purpose of the meeting is. This also applies to individual agenda items.
- Each agenda item should be allocated an approximate time for discussion.

- It is a mistake to permit members to read the minutes for the first time at the meeting. This wastes a lot of time if the minutes of the previous meeting are gone through in detail, item by item.
- Members should not be allowed to waffle, air their own prejudices or deviate from the issue being discussed.
- The chairperson should not use the meeting as a platform to air views.
- Recognise some members may have opposing views, and permit them time. Differences of opinion often produce the most beneficial results.
- Realise that it may not always be practical to come to a consensus.
- The chairperson (or secretary) needs to understand any constitution the meeting operates to.
- Ensure the chairperson can see everybody at the meeting.
- Avoid people that may conflict with each other from sitting side by side if possible.
- Ensure a quorum is present (25% of membership normally, unless laid down in the constitution differently).
- Ensure the name and the time of late arrivals and early leavers are noted in the minutes.
- Maintain order.
- Summarise the discussion of each agenda items; this helps the secretary.
- Close the meeting on time.

11.7.4 Informal meetings

These are meetings without formal agendas. They may be called, so preparation by the convenor can be carried out or may be instantaneous such as when walking the site and, for example, a sub-contractor makes an approach. In essence, the same rules apply as with the formal meetings less the formality and recording. It may however be necessary to confirm in writing any conclusions that have been agreed for purposes of the contract.

11.7.5 Telephone meetings and video-conferencing

Both are ideal for meetings where the members are long distances from each other and where to call a meeting would involve much travelling and associated cost. The telephone meetings work very simply. The convenor agrees with everybody a time for the meeting and at the appointed hour the telephone provider contacts the members in turn until all are connected and then the meeting can start.

The disadvantage of this form of meeting is the inability to see the other members and their reactions. It is difficult to maintain concentration if the meeting is lengthy and/or has too many agenda items. The latter especially if a member is not involved in most of the other items. So planning and sequence of items of the agenda is very important so that participants can drop out once their contribution is no longer required.

Video-conferencing is the same except you can see the person that is talking, but there can be a time delay in the sound. No doubt as the technologies improve, this will cease to be a problem.

11.8 Public speaking and presentations

The industry has increasingly moved in recent years to construction professionals making presentations to potential clients in an effort to obtain work or commissions.

11.8.1 Public speaking

There are four stages of oral public presentations: planning, preparing, practising and presenting.

Planning comprises three main parts:

- *Introduction*. Issues that need to be considered are: thanking the audience for allowing the company to present; stating who you are, who the presenters are and which sections they are to cover; the reasons for the presentation; explanations of any supporting documentation; when questions are permitted; and deciding on the duration of the presentation.
- *The main body*. The points to be made should be listed: these are then sorted into a logical structure, and a decision made as to the key points; decide the amount of time to be given to each; and determine what supporting documentation or visual aids are required.
- *Summary*. Which are the key points you wish to remind the audience of; and give the audience opportunity to question.

The next stage is to prepare. Trying to memorise the presentation is dangerous, especially in the event of stage fright on the day of the presentation. It is acceptable to use small note cards or similar as prompts. Any visuals used should be attractive, brief, clear, stress the important points, have large enough letters to be readily read by an audience with a selected font and should show the company logo.

Having established the duration of the presentation, practising will establish problems as well as give the presenters confidence in themselves and their colleagues. This is a good time to check if the presentation can be done in the allotted time. Ask colleagues to critique each other's presentation in terms of content and stance, oral skills, and so on.

Finally, give the presentation. Speak clearly; use pauses for effect and to allow the audience to digest; speak with enthusiasm and excitement; keep good eye contact with audience; smile and demonstrate you are pleased to be present.

Appearance is extremely important at any presentation. It is a demonstration of the speaker's standing and attitude, but also shows respect to the audience. There is a code of dress or convention appropriate to the occasion. If addressing a group of fellow construction professionals or the client, the dress code might be much more formal than if addressing a classroom of young children.

The body language of the presenter indicates much to the audience. No hands in pockets, keep eye contact with different members of the audience, use hands to make a point where appropriate, stand straight. These all give a positive image to the audience of confidence and enthusiasm. If it is a group presentation then those not involved at the time should show interest in what the speaker is saying even though they are only too aware of the content. They should never do anything to detract from the speaker.

11.9 Internal communications in organisations

The key to organisational communications is having clearly defined lines of communication with which everybody in the organisation is familiar. These lines of communications can vary depending on their nature. For example, the straight executive function as shown in Figure 11.2; or the functional lines of communication as shown in Figure 11.3; or in the case of safety another route will almost certainly have been produced as for instance in Figure 11.4.

In other words, the lines of communication do not necessarily reflect the organisation chart previously discussed in Chapter 2.

11.10 Types of internal communication

There are various types of communications which occur within the organisation.

- *Directives.* These come from the board of directors and are usually concerned with policy.

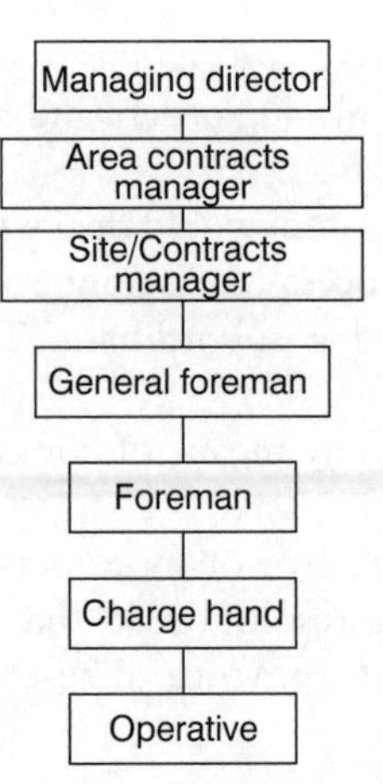

Figure 11.2 Executive function

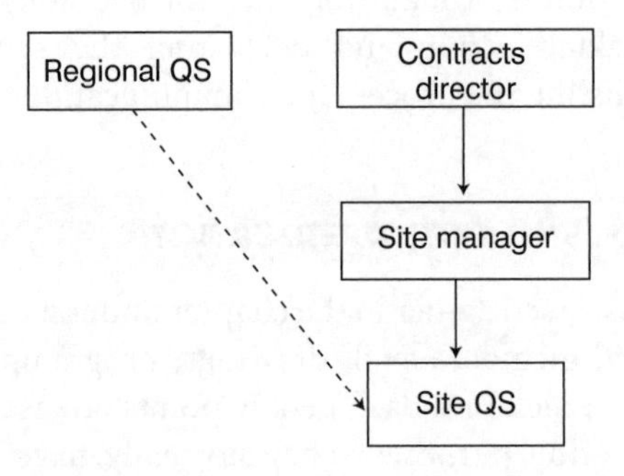

Figure 11.3 Functional communications

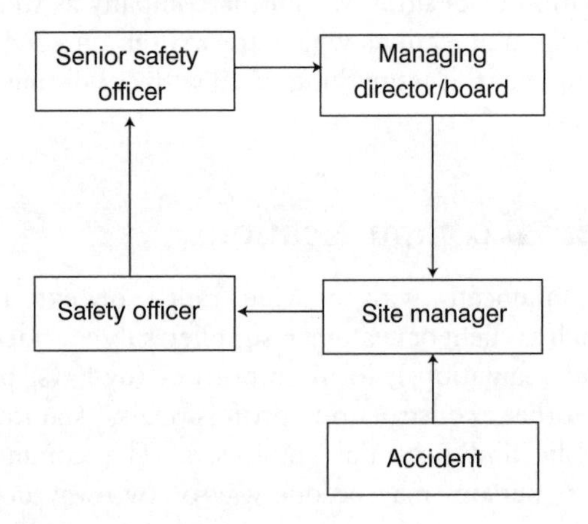

Figure 11.4 Communications for safety

- *Instructions and revisions to procedures or conditions.* For example, new types of paperwork for placing orders and revisions to employees' terms of contract.
- *Discipline.* The previous item would deal with grievance and disciplinary procedures, but the process of discipline is a communication between a member of staff and a subordinate. This is not necessarily their immediate superior.
- *Information.* The everyday means of communicating up and down the organisation.
- *Decision making.* Where two or more persons are brought together to discuss issues requiring resolution so that once the decision has been made it can be put into practice. These can either be structured or informal.
- *The structured regular meetings.* These can be weekly, monthly, quarterly or even annually. They normally have a standard agenda where information is brought together for decisions to be made about confirming, taking action and delegating the action and even on occasions discussing the process for communicating.

11.11 Methods of communication

These have been discussed earlier in the chapter under face-to-face and group communication and include: email, drawings, programmes and telephone. To this can be added memoranda. The key points are as a letter, but are for internal communication purposes. They normally have a standard format to include: To, From, Date and Subject. The first two may well have the person's designation or position within the company as well. There will be certain internal communications which are written on company letterhead; an obvious example of this is anything to do with conditions of employment of an individual.

11.12 External communications

Companies communicate with a wide range of external bodies and organisations such as client or customer, supplier, sub-contractor, government (for statutes and regulations), local authorities (by-laws, planning issues), trade unions, other construction professionals, statutory authorities, the general public and potential employees. The communications with outside bodies or persons may be one-way or two-way depending on the circumstances.

It is very important that the lines of communication between these various parties are clearly defined. If not there becomes a very rapid breakdown in communications. This is particularly noticeable when an external body starts contacting various people within an organisation, as the left hand does not know what the right hand is doing. It is important to know to whom drawings have to be sent and who has the authority to make decisions.

References

Davies, J.W. (2001) *Communication Skills: A Guide for Engineering and Applied Science Students*. Prentice-Hall.

Emmitt, S. and Gorse, C. (2003) *Construction Communication*. Blackwell Publishing.

Morris, D. (1999) [1967] *The Naked Ape*. Delta.

Kuhnke, E. (2007) *Body Language for Dummies*. John Wiley & Sons.

Index

The Death of Distance 2.0

How the Communications Revolution Will Change Our Lives

Frances Cairncross

First published in Great Britain in 2001 by

TEXERE Publishing Limited
71–77 Leadenhall Street
London EC3A 3DE
Tel: +44 (0)20 7204 3644
Fax: +44 (0)20 7208 6701
www.etexere.co.uk

A subsidiary of

TEXERE LLC
55 East 52nd Street
New York, NY 10055
Tel: +1 (212) 317 5106
Fax: +1 (212) 317 5178
www.etexere.com

A CIP catalogue record for this book is available from the British Library

ISBN 1-58799-089-X

Printed and bound in Great Britain by MPG Books, Bodmin, Cornwall